SMOKING

Smoking Red

The Red Arrows and more - a life on the wing

Wyndham Ward

YOUCAXTON
PUBLICATIONS

ISBN 978-1-914424-58-8

Published by YouCaxton Publications 2022

YouCaxton Publications

www.youcaxton.co.uk

For my Grandsons
Harry and Jack Konig

image © Michael Rondot collectair.co.uk

"Once you have tasted flight, you will forever walk the earth with your eyes skywards for there you have been, and there you will always long to return."
Leonardo de Vinci

The RAF is a meritocracy that brings out the finest in people and gives a chance to succeed to all those who respect and admire its traditions. This is a light-hearted account of a boy who was determined to fly whatever the odds. From horseback to cockpit. A journey that takes him from the green slopes of a hill farm in Wales to gain the coveted RAF pilot's wings in a proud service.

Preface

AFTER A week of watching aeroplanes almost collide and sometimes crash, it might be sensible to seek the serenity of the countryside to avoid a nervous twitch. But there's no need because these dramas are in a full-motion flight simulator anchored to the ground. And I love every minute since I am the one that blows engines up or sets fire to them to ruin a landing at London Gatwick. I can, with a flick of a finger summon a tornado over Brighton to make matters even worse.

In Montgomery, when friends encourage me to write about flying I have the perfect excuse because unlike many pilots I have never done anything remarkable. I've never been a hero or volunteered for NASA moon landings and claim in all honesty to have never made any distinguished contribution to aviation at all. Not ever. But my impressive under-achievement was ignored on a rainy night in my local watering hole, Monty's Cottage. The owner Russ and his wife Pam tasted a new beer with me while farmers talked about big tractors stuck axle-deep in mud. One remarked, 'I needed wings to get out of the mud today!' Eyes swivelled my way.

A sage suggested I write something for youngsters to convince them flying was better than driving a big John Deere unfairly pointing out I might have a few stories for my grandson. Harry's recent arrival made me a brand new grandfather but it was a bit early to think he might follow me. I hadn't met the little fellow yet – he was born in Australia. But too late, the idea propelled me back to my first 'two versus four' dogfight; into a cockpit smelling of sweat, burnt kerosene and old cordite:

Our four Hunters burst out of cloud into a world of eye-watering blue. Easing power off I slid into a fighting battle position on the wing searching for trouble. Squinting into the white orb of the sun is sweaty stuff. But hey! We had the advantage of numbers. No way could we lose. Suddenly two dark shapes dropped out of the sun curving wickedly into our rear for a gun's attack and I hadn't a clue where they came from. The fight was on!

Turning hard to face the threat I'm pinned in my seat with g-pants digging hard into my thighs to stop blacking out. Someone raps out a sighting call and we turn harder. Our Squadron Boss has bounced us and he was good. Too good, and in no time at all, aircraft were all over the place and we lost the fight and all on gunsight film for future humiliation.

I got lost. One second aircraft flashed past my head a hundred feet away then suddenly with the engine thundering I hurtle into a clear sky. I banked left and right to pick someone and get some sense of the fight.

I saw the Boss first – in my rear-view mirror with enough gunsight film to make me a Hollywood Star. Flying cheekily alongside he gave a jaunty wave then a drinking motion showing he was low on fuel. I slipped onto his wing trying to forget about the ribbing coming my way. God, it was fun! I put pen to paper.

Never expect a pilot to tell you why he wants to fly in less than twenty minutes because there is no scientific formula to measure enthusiasm and explaining a dream is never easy. Join me as a squadron 'Junior Joe' trying to shoot straight learning the art of aerial combat in a Hunter fighter. For more thrills, the Fleet Air Arm is hard to beat. Catapulting off *Ark Royal* is an unimaginably brisk way of going to work and a mind-blowing tail- hooking on is the most violent way of returning home, without killing yourself.

Flying is a discipline but can be entertaining especially flying in the Red Arrows. *Smoking Red* isn't written for my peers but for youngsters who aspire to fly whatever their circumstances. And to those fine people who follow aviation, I hope you enjoy this light-hearted view and laugh at my mistakes.

Many individuals named are sadly no longer with us, but, are mentioned in the present tense because as friends and comrades that's how they will always be remembered. Aircrews indulge in humour to dismiss danger and during my days we seldom had the political correctness we have now, so I hope none of the ribald comments used causes any offence. And if they do well …

Contents

Part 1

Chapter One

Country Boy with Gunpowder

EXACT POSITION, Tan-y-Glog Farm, Dolfor, Montgomeryshire County.

My unruly boyhood propelled me toward life as a pilot. This period was something out of 'Ripping Yarns' and it was a surprise to my parents that I survived boyhood, that our home wasn't razed to the ground and that any wildlife remained in our parish. They had a point because I accidentally burned most of our hill of three square miles of gorse, grass and not too many trees. Moving from Glamorgan my parents leased a small farm in Mid Wales named Tan-Y-Glog ('o' pronounced as in 'glowed') it sprawled under a steep hill with high stony crags. Buzzards wheeled overhead, a stunning place to live and the spectacular scenery attracted low-flying aeroplanes. Under the shadow of the same hill and confusingly called Glog Farm, our neighbour's yard opened onto its slopes. The hill divided lush pastures in the north and wild moorland to the south where bitterly cold winters and dangerous drifting snow meant ewes had to be herded down to safer grazing. A great place to grow up with my elder brother Eddie.

Glog Farm owned by amiable Julien Morgan and his wife Dora and their two sons Bryan and David was a second home. Along with Sally Dettore, daughter of our farm's owner we went to school two miles away in Dolfor where the teacher, Miss Easthope, turned not a hair if someone turned up on a pony and in bad weather pupils arrived on tractors perched like wet sparrows on a bale of straw. School was a room in a creaky church hall with wooden trestle tables warmed by a large pot-bellied coke stove. A sheepdog pup or two in attendance was normal but orphan lambs were discouraged because bleating interfered with reading lessons. In dark winter, the stove glowed a cheerful pink making our hall smell of damp wool, drying hair and wet dogs.

Textbooks appeared by magic and once a week a brown library van stopped outside to exchange books.

No upbringing in Wales is complete without chapel on a Sunday. There was no front pew for the landed gentry. Instead, the congregation by unstated agreement seated from front to rear in descending order of devotion, which put us, boys, at the back. Sermons were immensely long when Percy, a farmer and lay preacher was onto his favourite prophet, Elija. During these marathons, Bryan sat next to me and we could take a watch to bits with screwdrivers and reassemble it before Percy slammed his fist onto the bible to finish off Elija's enemies.

Before winter we sawed up fallen trees to feed the stoves, some trees were large to tackle so my father obtained a blasting licence for gun powder and within a day I knew where he kept the key to his powder store. One frosty morning near a fallen oak, l listened to my father's views on fuse length. Too short you blew yourself up. It had to be long enough to walk away tidily, not run like hell, trip up and get hit by flying wood.

Producing a pack of Swan Vesta matches, he lit the fuse with a flourish and we walked calmly away.

I think Dad was more used to dynamite and overcompensated with the gunpowder judging by the huge explosion. Large chunks of flying oak showered the pasture and the lingering smell of burned black powder did it for me! After secretly experimenting with powder from twelve bore cartridges and fine grains from .22 bullets mixed with 'borrowed' black powder, a winning combination emerged.

Wrapped in greaseproof paper with long fuses they looked brilliant.

I kept my illicit stock of bombs under my bed in a shoebox for safety until autumn. A rare sun dried out acres of grass and gorse covering the hill and our back garden stretched upwards to this tinderbox. Dotted with beech trees, it was more field than a garden. Mother had taken issue with a large shrubbery in the middle of it. This abomination to her advancing flower beds and rhododendrons had to go. And I knew how to do it. It was to be a surprise disappearance. When she was at the market.

My brother paled at the sight of my sack stuffed with gunpowder but although a bit uneasy about a strong wind I lit the fuse and walked slowly away.

Orange flames and thick acrid smoke surged away from the shrubbery and in minutes a crackling line of flames advanced uphill and I knew it was no longer a minor disaster; it was a full-blown one. The fire took hold and I suppose it was good luck for everyone that the thick cloud of smoke drifted overhead the only place which had a telephone. Fortunately, no livestock was lost thanks to the Fire Brigade and hard-working sheepdogs.

Despite a ban on explosions, life was fun and the day I decided to be a pilot began by saddling a pony to take a flock of ewes up the hill. Job over with a panting collie beside me I spurred up to the crag overlooking Tan y Glog Farm. It was a sunny day with a gentle breeze driving smells of warm heather. Overhead I heard curlews riding thermals in lazy arcs, their wings burnished gold by the sun and I envied their freedom in the sky. Perfect flying weather for them. And for others.

They flew from the east. A small shape grew into something I had seen in books. A biplane! It passed just below and I had a problem calming the pony because he had never seen an aeroplane either. I glimpsed two pilots wearing helmets with goggles looking up at me, with a wave they flew around the hill. Dragging the ponies head around to keep the aeroplane in view, *another* biplane flew past with two more men waving at me and it looked as if they were chasing the first one.

Galloping home, I informed everyone I was going to be a pilot. Had to be. No question. Bryan had seen them too, only he was on his farm much lower down so I was the obvious authority. Training started immediately. We needed a simulator. No problems. We built one. Born out of wooden packing cases and huge amounts of binder twine it was matchless. The site was a no-brainer: a massive oak in the plantation was perfect to hang it from and in no time we had a simulator complete with a wooden spade handle nailed on for a propeller.

To suspend our marvel, we pressed into service a thick rope from the granary winch, propulsion was no problem: a long length of binder twine tied to the base allowed us to drag it back like a swing. The rope snaked out and we found we could make it twirl by a pull to the side. Mistake. Rotating was a disaster after a meal so if we felt sick we eased off, and once vomiting stopped we started again.

Our parents were happy until I decided to think big and loop it. We needed more power in the swing and borrowing a tractor we lashed the pull rope to the front and reversed up the slope.

It was almost meticulous planning. But the floor was thin plywood and farmer Julian arrived to find his tractor just as Bryan released me. I dimly remember swooping up for a loop proper when the floor gave way and I was ejected through it. I was a bit dazed after hitting the ground and my arm was a mess from rusty nails. Father was summoned. We were grounded. So far the only casualty had been a heifer with a sore back that had the misfortune to be under our tree on my first, and only, parachute jump. The plan to convert our simulator into a bomber using gunpowder sticks to throw at targets would have to go on hold.

Grammar School flashed by. The Headmaster, Mr Lewis, a kindly man thought I should study veterinary at Aberystwyth but at fifteen, all I wanted to do was join the RAF to become a pilot. Everyone tried to change my mind but I made my way to RAF Cardigan, whizzed through tests and medicals and sat in front of an officer for an interview to join as a boy entrant.

He was a pilot. He asked me if I had any questions. I did. They poured out. I asked him what aircraft he had trained on? He said a Piston Provost, I wanted to know how it stalled, when did they teach you aerobatics and did he enjoy being a pilot? Politely he took back the interview but instead of rattling on about education, he asked me why I wanted to join. I think he regretted that. There was a lot for me to get through. Taking a deep breath I launched into the kind of career I wanted but he stopped me with a big grin. 'In January, you start at St Athan's and something tells me you will end up in a cockpit'. It was done. I was in.

I consoled my Headmaster I would take my O levels in the RAF and left with spectacularly little academic achievement. However, I unknowingly leave a mark: my athletic records for sprints and jumps remained unbroken for four decades. So I am told.

chapter Two

The Royal Air Force

A THIN grey drizzle shrouded the Montgomery hills as I travelled south to Cardiff by train. On the fifteenth of January 1961, I stepped down at a wet St Athan station clutching a small canvas bag to join the RAF. A figure moved alongside in a tweed jacket and country shoes like mine, sensing a kindred spirit I stuck out a hand.

'John Thorburn' he said with a grin'.

'Wyndham Ward' I responded, shaking his hand, striking up a friendship that lasted for life.

The next morning we were sworn in for a long time and hard-faced corporals started shouting at us immediately and never stopped for six weeks. There have been many stories of 'boot camp', and mine is no exception. A probationary period during our Initial Training Squadron, was smart planning by the RAF because they knew a thing or two about recruiting at our level. With an inventory of expensive aircraft, tons of inflammable fuel, almighty powerful bombs and rockets everywhere – there was every reason to be careful. Years of experience teaching youngsters to use submachine guns, rifles and bayonets, gave the RAF a remarkable insight into potential disasters.

Graduating from initial training was an outstanding achievement as nobody had been shot on the ranges or accidentally bayoneted. The RAF Regiment deserved credit for this, they were big on range discipline but had bags of flair. One of them showed us how to fire a Bren gun from the hip, Brilliant! Amazingly the RAF let me fire a machine gun aged fifteen but never trusted me to write home to Mum. The system didn't strip our character or chip away at our confidence as much as it instilled in us pride, discipline and a sense of purpose – a long-term view and it worked for me.

Squadron Leader Dunstan and his Stafford terrier welcomed us to our technical training hangar. Outside, big aircraft such as Canberra's and Britannia's stood ready. If ever there was fertile ground for something to go wrong, this was it. Teams of boys with hefty tools converged onto them dismantling various parts and running up engines with eye-gleaming enthusiasm. Anything could happen. And it did. But fortunately, the RAF had lots of engines.

Exceptionally fit-looking techs dropping by Dunstan's office. These were members of the RAF Mountain Rescue Team with a formidable reputation for toughness and Dunstan was in charge of the station team. Would we like to join? Mountain rescue life albeit in a junior role influenced my life, propelling me into a life of treks and wilderness overseas. Sometimes we took part in 'real rescues' where I proved almost useful. A wireless truck accompanied us and callouts were radioed into our base camp. This was manned by someone who could cook on a field cooker and one person who surpassed everybody was our boss Dunstan. One winter evening after training we looked forward to his roast chickens when we received a call out for a climber lost off Crib Goch Ridge and after a gruelling scramble through icy flooded causeways we found him. The Stafford Team assisted us to carry him down in the snow where the shivering climber's mate stood by a policeman. But incredibly his mate claimed he was the wrong bloke! Someone else was up there with life ebbing away in falling snow. Turning around we set off and amazingly the lead climbers found him and we brought him down. Thank God Dunstan had kept our chickens warm! It was my introduction to the RAF's familiarity between officers and ground crews; curiously this dividing line between us was never drawn in the sand, it went through our minds instead and it worked well.

chapter Three

The V-Force

RAF GAYDON in Warwickshire was home for a shorter time than expected. Victors and Valiants of the UK nuclear force were based there, guarded day and night by RAF armed police and fierce Alsatian dogs. I watched a white Victor start rolling the power of its engines made the ground tremble; but with every foot of runway used, I noticed the shape slipping from vulnerable to menacing. A formidable symbol of power.

The day arrived to report for work. Brilliant. I had finally arrived to do something useful and after an indecently clean technical school, it was paradise. Every battered door and wall was covered in oily handprints a smell of burnt kerosene lingered in the air and figures in oil-splattered 'cold wet' gear lounged in crew rooms drinking tea and cracking jokes. These were the line guys. Man's work! But probably a version of hell to the Station Warrant Officer a magnificent six-foot Polish pilot with a black silver-topped swagger stick who took a shine to me after I had the cheek to ask him about his flying.

An office dweller wearing corporal's stripes and a pompous face stumped up to me.

'Ward, follow me.'

We stopped at a door labelled 'Senior Engineer Officer' and I knew instantly something was wrong. I'd been confused with an experienced technician who knew something, better put him right. I saluted a smiling squadron leader behind a big desk, a picture of the Queen on the wall looked down at us, she was smiling too. I offered my name to clear any confusion but it seemed all ground crew were welcomed like this. However, at barely seventeen I was too young to sign for any job yet. The solution was to join the line mechanics and try to keep out of trouble.

I left pleased as punch and bounced off my new Line Chief. It felt like hitting a tree trunk. Master Technician Smith was formidable-looking. With a weather-beaten face and clad in a big parka he resembled a bear with warrant officer insignia on its paws. A wide canvas belt encircled his waist with ear defenders and a squawking radio hung from it which he ignored. Sticking out a hand he gave me a few welcome words.

'Get your ass into the crew room, Wardy, find Corporal Skinner and stick to him like bloody glue and don't get bloody run over by a bloody Victor.' He breezed into the office without even knocking and I realised instantly who ran the place.

My minder briefed me walking towards the menacing shape of a Victor crouched on the dispersal and I knew I would enjoy the bustle and roar of first-line servicing. Night scrambles on a rainy night were a theatre of flashing torches, dark shadows, flickering red beacons and the thunder of engines. The relieved chatter afterwards in the muggy warmth of the crew room, drinking hot tea, listening to crew chiefs was my introduction to a wonderful camaraderie.

The darker side of flying showed up the night of the 2nd of October. The crash alarm sounded and minutes later the first fire engine tore through a crash gate on the perimeter and roared across the country.

Victor XA 934 had taken off when the Number 4 engine exploded and caught fire. Number 3 was also shut down but during the approach, the Number One engine failed. The Navigator baled out but was killed on hitting a tree and the AEO died because it was too low for his 'chute to deploy. The first pilot was killed and the 2nd pilot ejected and survived.

The accident was barely over when the Cuban crisis arrived. My minor part in it was being hauled out of bed in the early hours to bomb up aircraft with live nuclear weapons. These bombs or 'buckets of sunshine' needed winching up into a bomb bay, which is easy. The difficult part was avoiding a bite by a truly fierce RAF Alsatian guard dog convinced I was stealing one. In the event the RAF was ready, JFK was ready, and the small bite in my backside was insignificant in the world order.

To sort out my education I went to night school and enjoyed a huge amount of sports. Every Wednesday, I fenced epee and sabre or played rugby. The RAF even sent me to a riding academy where a huge blonde top-heavy German woman in incredibly tight breeches strode around a ménage cracking a long whip. Devoid of one crumb of mercy she thundered at me to keep my back straight, toes in and taught me to jump properly; but never cured my appalling 'saddle slouch'.

Life bordered on ecstasy the day my pay of a guinea a week was due to increase seven-fold. Without a girlfriend or car, savings would be immense so I opened a bank account in anticipation of money pouring in. But it didn't and phoning accounts to accuse them of fraud was a mistake. A curt letter ordered a visit where an exasperated accounts clerk addressed me like a juvenile (technically I was).

'You have had six uniforms and six pairs of shoes. God only knows what else you have ruined, blown up and lost in equipment since you joined, so *you* owe the RAF money. You, my sunshine, are in debt to the Queen.'

After paying off my debt to the Crown my account showed a time of plenty and without the aid of a financial advisor, I found something to blow it on. A notice in the Air Force News invited participants to an Outward Bound Flying course at a private boarding school and the magic offer included instruction on the Thruxton Jackeroo and Tiger Moth at nearby Andover. Ignoring eye-watering costs in the small print, I wrote off immediately and applied for a student pilots' licence which I was barely old enough to hold. Leave was no problem. I never took any.

News arrived, I was the only person on the course and it was cancelled. However, I could stay with them and complete the flying at Thruxton at a reduced cost. Could I phone them?

There to greet me at Andover station was a chap called Charles dressed in a shapeless tweed jacket and baggy corduroys. Radiating an air of old-fashioned politeness he escorted me to a dusty old Humber and creaked our way to the school. Turning into a pot-holed drive flanked by a Vampire jet and a 25lb artillery piece we arrived at an old brick manor house with a faded air of past wealth where a skinny fifth former with what looked suspiciously like terminal acne grabbed my bag and showed me to my room.

Thruxton Aerodrome was near Andover. The Wiltshire Flying School were ensconced there for many years and the field housed an amazing little factory producing the Jackaroo which was designed around the Tiger Moth biplane. Featuring an enclosed cockpit that seated four, it came with a low price tag and claimed to be the cheapest four-seater in the world – and I was about to find out why.

A British racing-green supercharged Bentley with a driver clad in a flying jacket, scarf and goggles hurtled past trailing blue exhaust smoke. With a casual wave, the chief flying instructor, for it was he, left us far behind. This was civil flying!

The secretary was expecting me. She welcomed me with forms to sign that said that if I killed myself it was not their fault. My instructor was in the bar. Standing in the doorway I cautiously looked in, figures in flying overalls clustered around the bar, their conversations floating to me over the smells of frying bacon. A figure detached from the group and limped over.

'I'm Glenn your instructor' he said offering his hand.

A livid scar ran down one side of his face to his lower jaw and although I sensed I was in good hands I hoped the scar wasn't from crashing aeroplanes.

Glenn, I was to discover, was a patient, skilled instructor, who took enormous pleasure in teaching a novice like me to fly. He was a bit unorthodox in his view on discipline but loved a challenge. He was about to get one.

"How do you do sir, my name's Wyndham Ward" I said pitching my voice low to appear older.

'Bloody hell! Ward are you sure you are old enough to fly? He said and without waiting for my reply, limped into the locker room – 'Right follow me.'

Our briefing continued on the move, snatching up a leather helmet with dangling tubes, we made our way out towards the line of aeroplanes.

Difficult to sum up my feelings. Smells of high octane petrol, oil and rubber competed with the sight of silver biplane wings covered in flimsy fabric damp with dew.

‘Weathers bloody perfect’ Glenn remarked stopping at the first aircraft. ‘This is our kite, you sit in the front, I sit in the back. Get up on the wing and I’ll show you where everything is.’

He pointed out various controls and seemed pleased I knew what they were all for.

‘Next, our communications.’ He gestured for me to slip on the old leather flying helmet, two tubes from the earpieces joined a long tube that he draped over the back of my seat. Reaching into the rear he fished out a blue plastic funnel used to fill bottles and stuck it into the tube. I realised this was a homemade ‘Gosport speaking tube’ much cheaper than an electrical intercom.

‘Testing testing’ he growled softly into the funnel.

’Loud and clear’ I shouted. ‘How do I talk to you?’

’You don’t’. Just bloody listen.’ Communications were one way!

The Perspex canopy slammed closed. He shouted instructions through a small sliding window and I listened to the exchange.

‘Magnetos off’. A swing of the prop by the mechanic to prime the cylinders.

‘Clear, mags on’ a voice rumbled in my ear.

The engine fired up. Glenn chanted checks and we moved off to a flashing green from the tower. I expected to taxi to a runway but we bumped onto the grass and faced into the wind. He never wasted fuel. More checks; a steady green flashed from the tower and we were away. The 120-hp Gypsy Major engine roared the acceleration was faster than I expected and after bumping over clumps of tufted grass we lifted off. The engine noise through the thin Perspex- canopy and air whistling through the struts and wires sounded utterly marvellous and when the airfield dropped slowly away,’ I knew instantly this was for me.

Glenn’s calm tones in my helmet told me to follow through on the controls as he began a series of turns and short climbs and descents. He chatted about power, stick position and feel of the rudder, about the position of the nose with the horizon and how to ‘feel the aeroplane’ – ‘above all keep a good lookout’.

His instruction: ‘OK you have control[2] marked the start of my flying career and the highs and lows of learning to fly. We started with the lows as soon as I gripped the stick.

For a second nothing happened then noticing we were in a slight climb I pushed forward hard to correct, the nose went down sharply and I ballooned up hard against my straps. Horrified, I snatched the stick back again which shot the nose back up forcing me down into my seat again…

After a series of violent dolphin-like manoeuvres, Glenn wisely intervened before I tore the wings off. I released my stranglehold on the stick and felt it move gently – and suddenly we were flying smoothly without a whisker of height loss. I expected to see his head halfway through the Perspex canopy roof but he had probably given an extra tug on his straps when he handed over. I was distraught; I had read all the books – but I wasn't getting anywhere and felt low, desperately low.

My dream was in pieces until Glenn's casual tones filled my helmet. How he managed to convey knowledge and refined sarcasm without any insult was quite extraordinary. Later, I discovered this was the hallmark of a good instructor.

'You have a couple of basic faults, Ward.

Number one you hold the stick in a death grip and are far too heavy-handed. Fault Number two, you over-correct drastically when the nose moves an inch and for a moment there I thought you were starting a loop'.

'When I hand over you must hold the stick by the thumb and index finger gently, flying is a sensitive skill as befits a gentleman.'

Taking a deep breath, I did as I was told and discovered like countless aviators, it didn't require any effort at all to fly. A gentle correction was all it took and the aeroplane responded easily. I promptly did a few steep turns coordinating the rudder with the stick holding it gingerly as if it was a vicious ferret. My spirits shot up – I could do it!

Returning to the field, Glenn landed talking me through everything showing me how to taxi weaving the nose from side to side so that I could see ahead.

Jumping off the wing to join Glenn I was utterly speechless. My eyes must have betrayed my emotions; he gave a wink and roared with laughter.

'you'll do young Ward but we still have a lot of work ahead, now bugger off, see you same time tomorrow.'

I awoke to a glorious day for flying. Glenn met me with his usual grin and drew me into a small side room. At one end stood a blackboard on a rickety stand and on a wooden table to one side were dusty models of aeroplanes. He motioned me to a battered leather chair and talked about stalling and how to fly a circuit and what to do if he dropped dead in the air.

'OK let's have a look at side-slipping shall we. It's a good way to lose height rapidly if you make a cockup so I imagine you will be doing a lot of it.'

We finished with the 'gyroscopic' effect of the prop turning one way and how the torque causes one wheel to press harder into the ground than the other, making the aeroplane turn as it accelerates. I couldn't see any real problem but decided to keep quiet.

'Right – any questions?'

'You didn't mention when to brake on landing.'

'No problem with brakes.'

'Why's that sir?'

'We don't have any.'

I was trusted with the start which I managed without sending the mechanic cartwheeling through the air. A mild breeze flattened the grass and ahead of me two Jackaroos' taxied swinging their noses like me and I managed not to hit either of them. Suddenly I smelled burning from the rear of our aircraft. It was a vaguely familiar smell but mixed with the cockpit leather and aroma of dope it alarmed me. Twisting around I saw Glenn leaning back against a large 'No Smoking' sign – puffing away at a cigarette. He made a shooing motion and I got on with taxying. The other Jackaroos accelerate away in a straight line and frankly, it looked dead easy and I wondered what all this gyroscopic stuff was about. I lined up into the wind, Glenn's voice with a hint of cigarette smoke came through the Gosport tubes: 'smooth and slow with the throttle, remember the rudder, let the nose come up and gently on the stick.' A green flashed from the tower and I made my first take-off… Thruxton has a lot of grass between the runways which was just as well because my take-off was a total shambles and I can still remember with complete accuracy the startling amount of grass I used up. In my effort to impress

I gunned the engine, the tail came up rapidly and the torque slewed the nose around. Mesmerised by a horizon rapidly going sideways, I forgot the rudder. Until I felt Glenn's foot press hard against it. The horizon stopped, we straightened 90 degrees from our take-off heading and off came the power.

Glenn continued calmly 'Might want to try that again Wardy. Remember; rudder and power together. Off you go before we run out of grass.'

I was learning fast that if you did as you were told in aviation it helps enormously. Yawing a little until I got the rudder sorted we held a straight course, bumped across a runway, which came out of nowhere and lifted off without hitting the boundary fence. I felt a pulse of success at doing something right and climbed up to begin turns and stalls.

Glenn ordered me to level off.

'OK show me a steep turn to the right.'

With yesterday's revelation that gentle movements were required, I felt an ace. We demonstrated stalling. Encouraging me when I fouled up he prompted me with mild sarcasm to make it fun – until he asked me to set a course for home. I was stunned, I had no idea where we were. I was happy learning to fly without any navigation stuff. 'Not certain of where we are' I admitted. Glenn broke a long embarrassing pause.

'OK perhaps to the nearest county then.' He let this sink in. '

'Good airmanship means you know where you are at all times, if you were old enough to drink I'd fine you a beer!'

The slum of my first departure niggled me so a brilliant landing was called for. I couldn't see a problem; all I had to do was to put it down in one piece to impress everyone. The only problem was the last twenty feet. Not having any idea where the ground is, will always flush out a bad tailwheel landing, particularly if you stall the aeroplane from that height. We eventually came to a halt after four gut-wrenching bounces.

I fully expected Glenn to hurl me out of the Jackaroo and off the field. Instead, with remarkable sang-froid, he remarked: 'I think the fourth landing was the best, now have a good look at the front coaming and see where the horizon cuts it. If you can remember it we can skip all this kangaroo stuff. When the horizon reaches the same point trickle the power off. My old mum could do it Wardy.'

I learned something else about aviation: it's full of contradictions. Alert mind and relaxed touch, I was assured, would do it – and the last 20 feet were scary no more.

It just goes to show what a good instructor could do with an incompetent.

I pitched up the following morning to a smiling Glenn with spare overalls slung over his shoulder. He tossed them at me.

'Bloody good show Wardy. We've got the Tiger.'

I was taken aback at aeros so soon and he caught my look. 'Don't worry you are doing well you can take-off and land, stall, and turn without much height loss and besides I like your spirit.'

Our Tiger was lined up ready. Narrower than the Jackeroo but with the same engine it was renowned for its aerobatic qualities. Top speed was marginally better than the Jackeroo and I was determined not to crash this beautiful aeroplane. Scrambling up we spent time strapping in. Unlike the Jackaroo there was no canopy – just a small windscreen and the straps were broad canvas affairs that pinned you in. Everything was different with my head in the open, the engine noise was louder and I could smell the grass as we taxied to the take-off markers. I throttled up the engine roared, we accelerated quickly, the tail came up we bumped a couple of times and lifted off in a straight line and the blood flooding through my veins was probably faster than the fuel going through the engine fuel pump.

I could hear the wind whistling past struts and wires more clearly than the Jackeroo. I had the same leather helmet with the refined communications system but now I had to pull the goggles down when I poked my head over the side. A chilly slipstream buffeted my face and I laughed with the joy of it all. Glenn joined in with me. I think my enthusiasm touched a chord with him. I 'beat up' clouds and threw the Tiger around to get the feel of it. We stalled and he taught me how to get out of a spin and out of trouble.

Whatever I did must have been OK because he yelled: 'First loop Wardy' and dived to get airspeed before pulling up into a loop. I noticed he held the wings perfectly level over the top. It looked dead easy.

Overconfident, I fell out of two loops in a welter of g forces but got the third right. Barrel rolls followed without a twinge of airsickness (thanks to our spinning flight simulator in Dolfor) then I set a course for home. On the way, Glenn ran me through engine failure and practised forced landings in fields. I got cocky on one approach and in such a massive sideslip to lose height that I was almost at right angles looking down the wing. Suddenly everything went blue and I coughed furiously. A cloud of smoke poured out from my helmet.

'You might get a lot of smoke if the engine packs up' Glenn shouted as he took another big drag on his cigarette and blew the smoke down the Gosport tube into my ears. 'Don't get cocky Wardy.'

Flights flashed by and it was time to solo. But there was a problem – my Student Licence had not arrived from Gaydon. The Chief Pilot was non-plussed. 'No sweat Wardy we put Glenn in the back in a straight-jacket, can't log it solo but what the hell.'

In the event, Glenn sat smoking away in the back and never said a word. He clambered out shook my hand and ruffled my hair. It wasn't quite the ending I wanted but I achieved something that mattered hugely to me.

chapter Four

East Africa

LIFE PROPELLED me swiftly towards pilot selection, I could be called within months. But a chit in my mail brought catastrophe. I was placed on PWR, a Preliminary Warning Role meaning a posting overseas at short notice.

Our orderly-room corporal, who knew of my master plan of being a pilot and felt I didn't know my place in life, smugly advised that PWR ruled out applying for flying training. I would serve abroad and start again when I returned. Providing I was still in one piece of course. A colossal blow but ever optimistic I knew some postings were short because they were tough and others were long and a 'piece of cake'.

'OK, where am I going?' I was never happy with office dwellers.

'You'll know when you get there,' he said thrusting a list of postings at me and orders to report to the RAF Regiment to fire various weapons.

This could work out with the right choice. Some places were appealing, others not, where people shot at you, blew aircraft up and were pretty hostile. In these cases, I assumed the army would be nearby or perhaps a warship or two. I needed somewhere with a challenging climate, a short tour of duty and a quick return to my master plan. Pretty much faultless thinking.

Nothing in the Arctic or Antarctica but my hopes rose when I spotted Maralinga Range Australia. The notes said it was very outback, thousands of miles from anywhere and full of dangerous spiders, nasty snakes and drunken Australians. One-year tour. Easy number-one choice. I flicked past Aden, trouble brewing there with terrorist attacks and a two-year tour. Turning the page I spotted Gan, a blip in the Indian Ocean a one-year tour. Not an easy choice, Sarawak on the island of Borneo in the South China Sea looked interesting if there were any pirates left to bomb. Probably rained a lot though – so RAF Gan it was.

Satisfied, I shoved my choices at the corporal and reported to the Regiment range supervisor. A lean fit sun-tanned sergeant with a quiet intelligent air checked me off his list. There were five of us and I was the happiest. The others were married hoping for an accompanied tour to avoid domestic disaster. Using Lee-Enfield rifles we blazed away and as the others departed, the Regiment sergeant drew me aside.

'Hey lad, good shooting, like your attitude.'

Chatting over mugs of strong tea he filled me in on what was going on around the world and after answering tons of questions we fired .38' revolvers for some fun. A week later, I received a note to attend admin to pick up my posting. It was raining heavily and bursting impatiently into the place I leaned over the counter splashing water everywhere. The corporal ambled over smiling, which meant bad news.

'It's like this, Wardy, you're posted to a detachment of 30 Squadron Beverley aircraft in Kenya. Nobody knows when they leave or where they are going, it could be Aden, Haddraumatt, Kuwait or Muharraq.' He smiled. 'It will be somewhere lousy because it's unaccompanied.

'When' 'When?'

'Anytime this month.'

Free of line duties I reported to the Regiment to fire a Sterling Sub-machine gun, the friendly range sergeant pointed out there was trouble brewing in Kenya, might want to check it out. I made straight for the library, hauled down a political history of Africa, arrowed in on Kenya and checked the facts. There was trouble brewing, no mistake!The Northern Frontier District called NFD was a powder keg. Local nomadic Somali tribes kept moving deeper into Kenya murdering as many police as they could along with local Kenyans. Known as Shifta bandits, they aimed to unite the NFD with friendly Somalia, and my views of a friendly Kenya full of tourists faded when I realised I was going to be a target. Many ethnic Somalians in Kenya shared the Shifta view, one area housed so many the place was nicknamed 'Little Mogadishu'. Officially it was Eastleigh District and the government had concerns about insurgents lodging there.

So did I. The RAF base there was named RAF Eastleigh and guess where I was posted?

chapter Five

Kenya

AFTER TAKING anti-malaria tablets for a week I dropped my trousers in the sickbay where two medics with syringes took care of my health. One jabbed my left leg with Yellow Fever and Tetanus, the other did my right leg with Blackwater Fever, Typhus and a third injection of something I'd never heard of. Refusal wasn't an option. I would suffer a lingering death the minute I set foot in Africa without it.

Trooping out to East Africa in a civilian Brittannia from Stanstead, we refuelled at RAF Idris in Libya before arcing south over endless scrub and desert. I slept fitfully until the engines throttled back for our descent into Embakasi, Nairobi's airport, where the RAF based their jets.

For someone who hadn't ventured further east than Leamington Spa; the sights and smells of Africa with its heady mixture of earth and wood smoke are unforgettable. Jostling Africans wearing big smiles and colourful clothing lined the Arrival Hall and Exits.

Surfacing from a sleep full of expectations in my new world, I heard someone mutter about drums banging all night but I never heard them. A morning sun warmed the air filling it with smells of damp grass and red 'murram' earth. It's a distinctly African smell that stays with you always.

The RAF was handing over to the Kenya Air Force who were training their pilots and ground crews. The business end of the RAF was the other side of the base where 30 Squadron ground crews lived next to their Beverleys. 21 Squadron Scottish Twin Pioneers were closer to the tech site.; they could land almost anywhere in the bush and were highly active. The RAF site was closely guarded because the aircraft shipped out medical stores and ammunition to NFD and serviced stations from Swaziland to Aden. A harassed sergeant just off night guard duty briefed us on the situation. I couldn't wait for him to finish because I spotted a grimy notice of a flying club with a Tiger Moth and a safari club with a land rover.

With high hopes, I phoned the flying club.

'Is that the flying club?'

'Yes, man but Tiger did not come back so no club anymore,' said a mournful African voice. Deciding not to press the point I phoned the Safari club only to find someone had crashed something there too. A Cape-Buffalo head-butted their landrover up the backend and they had high hopes of fixing it once they had found it. It had been a long walk out of the bush for the driver and they were confident he could tell them where it was once he was well. I booked it.

Masses of barbed wire surrounded the station and in the distance smoke drifted up from straw huts crowding the wire. It was hot trudging around the perimeter towards the distinctive Beverleys parked there and I wondered what a new boy like me would do.

My daydreaming was interrupted by a screech of brakes. An open-topped land rover stopped alongside driven by the harassed-looking sergeant just off guard. Full marks though – he remembered my name.

'Get in Ward!' Great! a lift to the squadron, I made my way round to get in next to him. Wrong.

'No in the back,' he jerked a thumb towards the open rear.

'Lion on the bloody runway again.' This was new. So was the next bit.

'Full mag put one up the spout and keep the bloody safety on.' Keen to please I picked up a rifle lying on a bandolier of ammo and did as I was told. Bracing my legs apart and holding on to the roof gutter I managed to stay upright bumping over the grass. Going to work African style was something else.

Scanning ahead for a large cat I was enjoying a big game hunter moment when I saw a lion with a ruffed neck calmly walking on the grass alongside the runway. I yelled down and the driver leaned on the horn. No effect. Stopping at about 200 yards away, I unslung my rifle and shouted down to the cab.

'What do you want me to do?'

'Bloody shoot it! That's what! And do it when it's on the grass otherwise we have to lift it off the runway in case some aircraft bloody well hits it.'

Being a fan of simple instructions, I wrapped the sling around my arm to steady my aim and let fly – just as he moved off. Big bang. Miss. Puff of red dust alongside the lion, and wow did it move. But not as fast as the lioness next to it which I hadn't seen. The two lions bounded towards the fence disappeared in the long grass and appeared on the other side going like hell past thatched huts.

I assumed that in Africa people were used to seeing lions, but they were running everywhere. Job finished, I was dropped off where I had started instead of where I wanted to go and trudged towards the Squadron. I had already decided that being squeaky new I ought to be the quiet keen type and keep my mouth shut. It was a wise choice because I overheard a couple of army lads chatting about some stupid sod shooting at a lion instead of driving a 3-ton lorry at it, which was the normal way of dealing with them.

The squadron tech guys were welcoming. Even the Beverly aircraft parked on the dispersal looked friendly. I was shown my accommodation hut with white 'mosquito nets' hung over beds. With showers next door and a stone-built tea crew room next to the aircraft, what more could I ask for? Joseph, our tea steward, welcomed me with a smile and a brew. It was a neat setup and close to hand we had our own small armoury manned by a smiling Kenyan. It proved handy one morning when I almost stepped on a massive Puff Adder on the tea room steps. They are big ugly things and particularly venomous so I banged on the armoury door for a shotgun and a couple of rounds to sort the sod out. But when I rushed back, Joseph had hurled boiling water at it and two of his chums had beaten it to a messy pulp with long-handled spades. I loved Africa!

An enthusiastic Corporal John Pearson took me to the aircraft and on our way, we walked past a small building with a large grill on the front. I was intrigued when two small hairy olive arms rose from nowhere to grip the bars and a baboon hauled itself into view. John remarked her name was Jane, taken in as an orphan by the squadron. Jane and I became friends and our companionship caused a whole lot of trouble.

Beverleys were interesting. Slab-sided, powered by four Bristol Centaurus engines with large propellers and low-pressure tyres they could land just about anywhere. Rumour was that it was designed as a block of flats by the Air Ministry but the RAF liked it and asked for wings on it.

Engine fitters thought it, 'an ugly sod with engines burning as much oil as petrol'. But I loved the brutes; they were uncomplicated. The boom had a large hatch in the floor for dropping troops out. This was situated right next to the toilet which wasn't a good idea. After the first person killed himself stepping out of the toilet and falling through the hatch, pins were fitted so the toilet doors locked whenever the hatch was open.

John glanced skywards. 'Watch that one landing .'

Runway ends were black tarmac with a middle section of graded red murram earth. A huge shape loomed out of a blue sky landing on the tarmac, seconds later it hit murram, the huge props went into reverse pitch and the aircraft disappeared in a cloud of red dust. With a squeal of brakes, a massive shape emerged like a bull elephant out of a dust cloud and lumbered towards us. Astonishing. With a belch of oily smoke a Bev started up and its departure was as eye-catching as the one landing. A marshaller casually slid sunglasses on to avoid getting his eyes blasted with flying grit and motioned the Bev backwards. The props went into reverse pitch and like a lorry reversing out of a parking spot, it backed out, stopped, and then went forwards.

My first day ended with a huge red sun sinking below a wide horizon. The distant Ngong Hills slipped into purple shadow and with unnatural suddenness it was dark. Once night claimed our bit of Africa the evening chorus started up. Noisy crickets, loud screeches and baritone frogs sounded like an alien but friendly symphony. But that changed when I drew my mosquito net around me. Drums started beating across the wire. Now and then there was savagery to it and I wondered if this was a message to other drummers. In the end, they sent me to sleep and the pounding rhythm of nightly drums with the smell of red soil became comfortingly 'Africa'.

My new job was to start a Houchin engine-driven generator on wheels and haul it to an aircraft with a land rover. By this time I was friends with Jane and she particularly liked hopping up onto my Houchins to warm her backside near the exhaust. It was a small step from there to accompanying me onto the aircraft with a long line of cargo tape tied to her neck to watch me work.

Queens Regulations didn't forbid baboons accompanying me but I felt discretion was required, venturing into the cargo bay onely late at night or pre-dawn with few people around. I miss-timed it once which created a problem. I usually stashed a few nuts for her in my tool bag along with my screwdrivers and spanners and Jane had become adept at handing me tools. Dawn was breaking when I checked the wiring for a heavy drop and she was about to hand me a screwdriver when a tough old Flight Sergeant with leather features clumped up the ramp with a young para-officer.

'This one's for heavy drop sir and we are in luck because the electrician's here to explain the wiring.'

I scrambled up, Jane bounded up clutching my screwdriver and we stood facing them. 'I see,' said the para eying us both.

'And which one would be the electrician?'

I managed a weak smile before dragging Jane out and tying her to the Houchin followed by a stony-faced SNCO who demanded to know if I had taught her to use a hacksaw and wire cutters yet? If so, we were in trouble with baboons never mind the bloody terrorists and I was a bloody disgrace to the whole air force and bloody well going to hear more about this.

He may well have had a point about the wire cutters and I never took a baboon onto an aircraft again.

An even lower point in Ward-Kenyan relations emerged later when I took her to Nairobi and got banned from the famous Thorn Tree Café where she attacked a passing tourist. Or more accurately the poodle she was carrying. Baboons hate little dogs.

Not really my fault. I had tied her securely to a table leg and placed her discretely under the table with some nuts. Trouble came when Jane spotted the poodle and hurled herself at it with some sort of monkey war cry taking the table with her and scattering crockery everywhere. The head waiter got upset. I can't blame him although he narrowly missed a nasty bite after throwing pieces of expensive Dutch porcelain at her which she rightly objected to.

Life improved when I noticed horse droppings near a dusty track close to the squadron. The track to a paddock where I counted thirteen horses of mixed breed and size, all in good condition.

A short bronzed well-built chap with the hairiest body I ever clapped eyes on hailed me. 'Morning and who the hell are you?'

After this pleasantry, I introduced myself pointing toward the horses.

'Who owns these?' 'Army does. My name's Pete.'

He relaxed when he knew I was RAF. His regiment had left their horses behind when they returned to the UK. Instead of turning them loose, they left him on a posting as a groom which I thought was decent, ignoring the feeling the brightest pixie in the woods wouldn't be left behind.

'Who rides them?'

Nobody, do you ride?'

'You bet.' I replied nonchalantly.

'Which one do you want?' he asked sweeping a hairy hand at them.

A difficult choice as they were all sound. One large chestnut mare about seventeen three hands had long scars down one rump. He saw me eying them.

'Lion' he remarked. Phew! In my opinion that was the one horse in Africa that would smell a lion a mile off and a wise choice. But as I wasn't planning on meeting one I settled for a plucky-looking grey Arab gelding.

'That's Tomahawk, I'll saddle up.'

It was surreal. I was in Africa being offered a choice of horses and a groom tacking up for me. I sprinted back to my hut for breeches and joddy boots.

I mounted up to sketchy directions on where to go in the bush. He spilt out a jumbled description using Swahili names to avoid carnivores and a village best avoided on account of their politics. It was confusing so I decided to stop at our squadron armoury and sign out a .38', ideal for making noises to scare animals and cater for extreme political views it fitted nicely into my saddlebag. Brimming with confidence, I set off for a side gate via a long stretch of grass past the officer's mess over which Tomahawk cantered beautifully. Enjoying the warm sun on my back and the smell of strange grass I approached two friendly Askari (guards) who swung the gates open for me. With a wave, I trotted through in a clean shirt and breeches trying to give the impression I knew what I was doing in Africa.

I reined in to view. Marvellous! Flat-topped thorn trees dotted large yellow tracts of grass shimmering in the heat. Vultures perched on trees I couldn't recognise and spiky sisal plants sprinkled the bush to the horizon. With a goatskin of water slung over my saddle and enthusiasm in spades, I trotted on and in no time came on herds of wildebeest with shaggy necks and curled horns. Scanning the bush I noticed small knots of giraffe and gazelle with their heads down which meant they felt safe. And in no time at all I became overconfident.

Tomahawk was up for anything and I foolishly trusted him right up until I decided to get closer to a large herd of wildebeest. Touching my heels in to urge him forward was a mistake because the next minute we were at full gallop hurtling into the herd with me frantically sawing at his bit.

Pumped up and totally out of control it was obvious to me that Tomahawk had a thing about stampeding animals. The herd took off in grand style. It was exciting stuff with dodgy moments hurtling past tossing heads and wicked curved horns; flecks of saliva flew everywhere and the drumming of so many hooves on hard earth was awesome.

I have to admit there was a thread of panic too, which I ought to own up to but at that moment, I could have been with Crazy Horse at Little Big Horn or with the British Cavalry in Sudan. But the trouble with so many hooves churning up dust meant I couldn't see where I was going; Tomahawk didn't care so I held a straight course which was smart thinking on my part to get the shortest route through.

Suddenly we were into clear air. So clear I could see every detail of the dozen giraffe less than fifty feet away that I was going to collide with. The problem with surprising giraffes, I discovered, is their heads are above the dust and mine wasn't. They may have a graceful motion in a straight line but a rider bursting out of the dust makes them shy away in a welter of huge limbs and long necks. This can upset a horse because giraffes have a powerful kick.

However, Tomahawk was no fool: Arab horses can turn on a penny piece. Which he did. Lucky for me the giraffe was still feet away when Tomahawk turned hard. I went straight on narrowingly missing the beast and flew through the top of a small prickly thorn bush landing hard on sun-baked ground.

Remounting took ages because Tomahawk kept trotting out of reach, consequently, the sun was about to set when I approached the gate where both Askari's gaped at my ripped shirt, stained breeches and blood-streaked arms. Tomahawk unharmed and in my view, a bit cocky trotted through nicely with his tail up which hopefully made up for my appearance.

Getting thrown off in the bush close to nightfall was a concern so I asked if anyone would ride with me? Mistake. A mad Scot who claimed he rode a few racehorses in his time, seemed suitable and I was easily gulled. Bob was as wild as hell in the saddle and great company provided he stopped carrying whiskey in his saddlebags. Our agreement got off to a shaky start when he produced a large leather hunting flask he claimed was a clan tradition. One morning he consumed an entire flask before we rode out. I think he got a bit mixed up when we jumped the hedge near the officer's mess because he ended up on the veranda in a welter of tables, silver cutlery, and crisply folded napkins. It wasn't a pretty sight but eventually, we got it right, saddling up sober before dawn to ride in the bush, trotting in just before night fell.

Climbing Kilimanjaro was fun. Using three-ton lorries trailing red dust we camped in Arusha. On the mountain, I had to escort a couple of our guys down who looked ill just before the summit; the Lead asked me to take them to the saddle between Mwenzie the opposite peak. They were fine, so I went up again constantly short of breath but the view was awesome. Pity my camera didn't work.

Flying operations intensified. Main personnel left but guard duties had to be covered and some were active with armed thieves cutting the wire. Any parties in married quarters had an armed guard on the roof and guests were advised to carry a sidearm. NFD troubles were getting worse and rumours of leaving swept the squadron. Eventually, I delivered Jane to Nairobi Game Reserve Orphanage and said my goodbyes. Boarding an Argosy for Aden I left Africa with rich memories knowing the easy part of my plan was over, all I had to do now was complete the unpleasant part where people were more hostile and blew things up in a disagreeable climate.

chapter Six

Inconvenient Insurgents

ADEN WAS a Crown Colony and the Radfan hinterland part of the Aden Protectorate. With Britain in control and bringing peace to the Protectorate, there was revenue to be had but unfortunately, many armed tribesmen didn't see the fiscal side. The invention of cheap Japanese transistor radios, easy to get hold of, made Nasser's views on anti-western matters easy listening. Trouble was brewing.

Landing in Aden to a mounting escalation of violence any doubts about the situation were dispelled when we loaded onto a bus to our accommodation. Every window was covered in steel grills to prevent grenades from exploding inside; two watchful armed army guys positioned at either end next to the doors constantly scanned the rooftops – it was serious. The next morning, I found I had slept through a fusillade of gunshots from Crater the old Arab quarter which didn't say much for my alertness.

Dawn saw us on our Argosy bound for Muharraq via RAF Riyan, an isolated staging airfield 175 miles northeast on the coast of Hadhramaut, the wild feudal territory to the north of the Protectorate (which in a few years became South Yemen) The plan was to drop a load there and Salalah in Oman before heading north.

After a meal of dry curled-up paste sandwiches, we descended onto the coastal airstrip of Riyan over a landscape of raw high peaks split with deep wadis. It looked savagely beautiful but I imagined life must have been cruelly hard there.

I felt excited about Muscat and Oman. Not much was known about the place. I read explorers' accounts and filched RAF charts to get a feel for it. Described as a closed shuttered place with sketchy borders with large areas marked enticingly 'unsurveyed' it attracted me. Disembarking I had an eerie feeling of a future here. A Movements guy hurried after me warning me not to stray on account of land mines. Behind him was a small tough-looking tribesman holding an ancient Martini action rifle, his chest crossed with a bandolier. He trotted nimbly towards us looking comfortable in loose-fitting robes, his smile friendly. 'Askari' he announced and promptly faced the desert to indicate he was keeping us safe. My first glimpse of a slave in the Sultan's Army.

Disembarking at Muharraq, the flinty gaze of a tough-looking Regiment type fell on me. He directed me to one side to join several airmen who were roughly the same height. His words 'You lot will report for riot squad when called' held an ominous tone. It was late 1964 and they had reason to be cautious; a few years earlier a Beverley had been blown up yards from where we stood. A pencil bomb inserted in the refuelling line was the cause, three weeks or so later explosions followed in the storage area. Not so nice people here then.

Our Beverleys parked in the open on a hard scratty desert floor and the sand and coral dust caused problems when we marshalled aircraft backwards. In reverse pitch, the prop wash sandblasted faces, eyes and everything else so we wore goggles and Arab headdresses. I became established as a 'flying spanner' whenever the chance arose to accompany aircraft down route where engine mechs and sparkies were always handy. The squadron dropped 3 Para regularly who were a tough bunch. Operations including night drops involved swift work on the first Bev back so that it was ready to return to a secured strip to evacuate any casualties; some had bad injuries of broken ankles and backs. Heat and humidity increased daily and we sweated performing even minor tasks in aircraft exposed to the sun all day. Tools were covered with a rag to avoid burns and changing a propeller with rudimentary lifting gear in the middle of the day was particularly difficult. The whole task was made even worse when dust devils – miniature tornados of whirling sand – spiralled across the desert floor covering us in coral dust.

Guard duties became irksome as the political temperature mounted. As a flying spanner down-route into Muscat and Oman through rugged Hadhramaut to Aden, I found the atmosphere tense. Guard duties were not good. A land mine planted on the runway at Azaiba in Oman exploded under a landrover, a technician was killed. Viewing the charred wreckage in the morning I realised that quite easily could have happened to me.

It was a political tinder box from Bahrain down to Aden and isolated RAF stations down route were vulnerable. This was the situation in boring Muharraq early in 1965. A British withdrawal from Aden was the subject of open bar discussion so it seemed factions were sorting out who got the turf. My only concern was the rising tensions coincided with exams I needed to pass for pilot selection. Ever optimistic I hoped that apart from a surge of minor pilfering everything would blow over; it was a mystery why locals would steal all the glass sulphuric acid canisters from our fire extinguishers. But, steal them they did. And suddenly life wasn't boring anymore.

A riot squad formed, we kitted out with gas masks, webbing, a rifle and bayonet plus bandoliers of ammo and presented ourselves to 3 Para. In a cordoned-off area away from prying eyes, a large crowd of evil-looking types in ragged Arab dress were busy screaming at a 'U' shaped squad of paras armed with rifles who patiently ignored pelted stones, pieces of wood and howling insults. Inside the armed formation, a magistrate in flowing robes stood with an armed Bahraini policeman, a photographer plus two stretcher-men. Abruptly, the situation changed, The magistrate raised his bullhorn booming Arabic at the mob; policemen, flanked by armed paras marched forward, unreeled a tape onto the sand and sprinted back. A barked order: paras fixed bayonets, steel blades flashing wickedly in the sun. Impressive. Another order, half donned masks, and suddenly everyone was masked facing the howling mob. Insults mounted, the mob surged forward, the front paras went down on one knee aiming at one person, tear gas canisters arced upwards and as a dirty foot crossed the line a solitary shot rang out, and a body slumped to the ground. Before you could blink, the paras advanced over it halting smartly facing down the mob. The photographer snapped the body and the stretcher-bearers rolled him onto the stretcher. All done in less than a minute hidden from the crowd by the wall of paras. Not finished yet. Down went the paras again on one knee and aimed at another prominent individual. The mob promptly ran away giving V signs to their para mates and the demo was over.

We practised riot squad duties daily sweating in the heat and although we weren't paras, we were getting good. Whenever a siren wailed we stopped working, grabbed our rifles and kit and launched out in 3-tonners.

BAPCO the Bahrain Oil Company lit the spark that ignited the tinder box by sacking a large number of locals. Violence erupted. It was the beginning of a month-long Arab Intifilada with international repercussions that left people dead. The insurgency started with student demonstrations in Manama High School rapidly escalating with hard nuts infiltrating the mobs. And guess where most of these thugs were busy inciting howling protesters. Muharraq!

For a fleeting moment, the awful implication was insurgents might have got hold of my exam timetable and were out to ruin my plans. The feeling passed as soon as we faced the mob. It was hot wearing a steel helmet and a gas mask and although the comfortable feel of a loaded rifle and a fixed bayonet helped, the noise and implied violence were a bit unsettling, to be honest. In terms of academic preparation, it wasn't the best thing before an exam either. Particularly unnerving was the ululating noise from the women. There's savagery to it but strangely, the chanting: 'Down! Down! Colonialists' sounded comical and didn't faze me at all.

The good news was our stolen acid bottles turned up. The bad news was that they looked like grenades when they hurled them at us. Corner riflemen facing inwards to cover roofs, can by international law, fire independently if threatened and gas has been thrown. Fortunately, we were restrained about the situation but the police weren't and were prepared to fire. Sensing the switch the mob vanished into side alleys yelling threats which was a relief. I was already late for my exam.

Stinking of burnt rubber, sweat and rubbing itchy eyes on account of tear gas on my hands I presented myself to a bored-looking invigilating officer. Checking a list he snapped, 'You're late Ward, I am docking half an hour off the exam time.' But I was too busy finding room on the floor to put my rifle and smelly kit to care. Getting to be a pilot was hard work!

Life was not brilliant with frequent guard duties and riot call-outs. One early morning, bored witless after night guard duty I was strolling back to quarters when I heard a commotion outside a mudbrick building. A police dog had chased someone in and his handler was considering possibilities. I offered to put a few rounds through the door and walls if it helped. He looked at me as if I was out of my mind and told me where I could stuff my rifle and the last I saw of him he was firing his pistol at someone scuttling over the roof with his dog frantically trying to follow.

I don't think it's an overstatement to say I was pleased to pass my exams despite a large number of hostile people trying to stop me. Life was so good that on my way to admin' I didn't notice the ripe smell of the open sewer on the beach next door – it was called Turd Point for good reason. Tanned, fit, and ready for anything I was prepared to dismember any of the office tribe at the faintest hint of sarcasm. Didn't need to. It was tidings of great joy. The posting notice was blunt, report to 30 Maintenance Unit RAF Sealand in Cheshire in two weeks. Two weeks off was a bit mean but I had earned it!

chapter Seven

A Step Forward

FATHER GENTLY prised out the facts of my sojourn in Africa and Arabia: my letters home vaguely suggested fun and games under a warm sun meeting interesting people which to a point was correct. I enjoyed RAF Sealand, dipping my toes into the academic world at Liverpool University's open classes but on finding I passed my exams early, I ditched academia and applied for aircrew the instant the result was in my hand.

Rumour was the aircrew selection engine was well ahead of the training machine but notice came through with the ink barely dry on my application.

Pitching up at Biggin Hill for selection, naïve to the fact that thousands applied but only a few passed wasn't arrogance or presumption, I just felt my strong desire to become a pilot was bound to show through. It did. And so did arriving without any preparation.

I was super fit, passed the medical and sat various papers and calculations. My fellow candidates were a mixed bunch, varying from cheerfully blank to academic genius without common sense and most had prepared well.

Chewing the end of a pencil I gazed at the colossal number of questions on speed distance calculations. Must be a shortcut. Yup. Got it! At the finish, a host of bright sparks swaggered up to the invigilating officer while the remainder of us sat like stunned mullets.

Ok, bring on the next bit. Sadly not for some. Guys were falling by the way. Not so good.

A good night's rest then coordination tests with rotating gizmos and blinking lights. After resting eyeballs, it was a leadership assessment over obstacles with a team using planks too short, ropes not long enough and a few scattered barrels. Compared to the equipment I used in the heat, flies and dust, no problem. Next one.

Problem-solving in a hypothetical situation. Persuading a few stubborn individuals I had the misfortune to work with, incapable of listening to any sort of reason, made this easy. A measure of relief, my name is still on a list.

The next day. Relief vanished. Crunch time. Interviewed by a panel. One wore a navigator brevet the other was a pilot.

Good. They were flyers, not psychiatrists with a degree in flushing out people like me. I respected them but I wasn't fooled either, I had to convince them in a short time of aspirations held for a long time. They were pleasant but not over-cordial and I hoped my new flecked tweed suit by Gieves and Hawkes would give a good impression.

With my service record in front of them and being too young and clueless to have done anything of any value, it was never going to be a long interview. My entire life was summed up in minutes with academic achievements taking even less time. Nothing remotely interesting in the personal locker except flying Tiger moths and Jackaroos and while labouring this with an excessive zeal I noticed the Navigator's eyes glazing over. He seized the moment in a gap between heroic cross-wind landings and bowled a curved ball.

'So, what's most important to you Ward, being an officer or being a pilot?'

Bit unfair that one. 'Both are important' I replied. My immediate thought was, 'this chaps a terrier'. With a terrier, you belt them around the ears to sort them out but it's not a selection option. He narrowed the choice.

'Is it better for you to be a good officer then?'

It was definitely 's**t or bust' time now.

'If I may say so I would like to be a good officer and I could only be that if I were happy in what I was doing,' I plunging on.

'Being a pilot would mean just that and I would not be happy in any other role,' I paused. The line between diplomacy and interview lunacy hung in a delicate balance.

'As much as I respect other flyers.'

Silence.

After a shuffle of feet, the terrier dominated the food bowl completely.

'So, if offered the enviable role of Navigator as a commissioned officer, you would refuse?'

'Quite so,' I replied respectfully, firmly folding my arms.

I am sure the pilot on the board suppressed a smile when he stood signalling it was over. He told me to report back the next day before I left for Sealand. That was unusual and anything unusual in the RAF was never good news.

The next morning, I met the same officer. He shook my hand.

'You've done well Ward jumped all the hurdles and passed the aptitude tests for pilot training. But there's a hell of a backlog I'm afraid. The choice is yours. We're short of Air Electronics Operators and if you accept now, you can be a sergeant in weeks on full pay and we'll pull you off after a squadron tour for pilot training.'

'On the other hand, you start officer training in six months and be part of a backlog for at least three years.' I was bowled over at passing but felt diminished about having to wait so long. But I took his word and decided on the spot. He grinned, 'I couldn't see you as an officer in charge of bedding stores' (the RAF was to live up to its word and surprised everyone except me)

Packing took no time, the meagre boot space of my old TR2 was designed for a lightly drilled toothbrush and not much else. Two outstanding friends, Peter Egginton, whom I played rugby with and Matt Millan who I climbed with, were delighted for me. Our farewell beers lasted several days in the Boathouse Inn Chester and one fine morning with my shotgun in the back, cartridges in the boot and my only suitcase on the seat, I set off for RAF Topcliffe in Yorkshire, home of the RAF Electronics College. My foot on the flying ladder at last.

It turned out to be a hard but happy course. learned morse, listened to long lectures and played good rugby . Evenings were pleasant affairs listening to Mahler and Tchaikovski while shooting pigeons, hares and rabbits out of my window for the brilliant mess chefs.

chapter Eight

The Kipper Fleet & The Old Grey Lady

The old grey lady

EXACT POSITION rowdy end of RAF ST Mawgan Sergeants Mess.

In 1967 my time in Coastal Command started well because the Sergeant's Mess at St Mawgan had the reputation of being the best nightclub in Newquay. The atmosphere was upbeat and a friendly group of aircrew led by a charismatic South African Rex Wickins made me welcome. Compared to the V-force, Coastal Command's contribution to the cold war was entirely different. Here daily contact with the Soviets was a way of life where surface forces and subs were counted, shadowed and reported on the enemy reporting network.

Coastal Command in the UK was split into two groups: 18, in the north and 19 in the south and both had a western limit of 30 degrees west where the Americans and Canadians took over. The aircraft used was the Avro Shackleton sometimes referred to as the 'Old Grey Lady' or twenty thousand loose rivets in tight formation. It was a beast with a lot of character.

Powered by four Rolls Griffon engines with contra-rotating props it sounded marvellous – but inside it was noisy. Modifications allowed the carriage of nuclear depth bombs. Mark 2s had a tail wheel and Mark 3s had a nose wheel which was heavy and made it a handful with an engine failure.

Despite being young and wet having just left Electronics College I was probably the most enthusiastic of baby kippers. The quality of instruction was good, heavily laced with humour and I particularly liked target recognition identifying anything Soviet that floated or flew. Radar training was new and during Sonar we listened to weird sounds underwater. Shoals of snapping shrimps sounded like a distant thunderstorm raining on a tin roof. Squeaky dolphins chattered into our underwater mics and a high-pitched lonely call with a haunting quality was a whale talking to another whale. The lesson ended with a long deep reverberation like an underwater foghorn which almost blew my headset off: a calm voice in my earpiece said, 'That's a whale farting which means a coffee break.'

We put our training together in a simulator called a 'Stage 2' trainer. Experienced guys run it with a humorous approach that everyone was grateful for because things could get heated. The two navs triangulate our inputs with little beams of light on their plotting table and directed attacks with a moving aircraft symbol which was complicated enough without us baby-kippers screwing up sonar plots.

Without coming to blows as a crew we met the pilots and the T4 Shackleton. Fitted with extra radar positions for training it could remain airborne for a very long time which meant a lot of rations.

A Maritime Air Instructor (MAI) trained us as a fighting crew and Screen AEOs and Navigators mentored us. All were highly experienced and nothing seemed to worry them. There are many descriptions of climbing on board a Shack for the first time. I favoured: 'Bit like sticking your head up an elephant's bum, it's dark, smelly, and you don't know what you are going to find up there'. The T4 was a taildragger, so everything was uphill after you got on. Discounting an Elsan toilet near the door, the Shack smell was unique. An aroma of leather seats, hydraulic fluid, engine oil and galley smells greets a 10-man crew. High stepping over the main spar, two pilots, two navs, a wireless op and a flight engineer moved forward. Aft of the spar, were sonar, radar and two beam lookout seats. Opposite the galley, two rest bunks and to the rear banks of flare dischargers lined the side. A flame float dispenser chute opened in the floor and a six-barrelled photoflash discharger gleamed in the roof and two concave beam windows opened inboard for photography.

Starting was a noisy affair with the occasional belch of smoke and backfire but when the Griffons fired up, they idled with an unmistakable pleasant growl.

However, the RAF didn't pussyfoot about with soundproofing and insulation from the cold in our T4's. No sir. It was a man's life in the old Kipper Fleet. On take-off, the thundering of four supercharged engines at full power boosted with water-methanol injection positioned mere feet away was loud enough. But combined with a rising crescendo of rain pelting against the thin skin of the fuselage as we gained speed, the noise rose to awesome.

Throttling back after take-off reduced the noise and we used a conference intercom to chat or just shouted but either way, we started our journey to developing 'Shackleton Ear' and lip-reading. Winter with low-pressure systems romping up the western approaches was a good introduction to Coastal. For the first few hours, it was a matter of getting used to being thrown around the sky in foul weather; I was manning a beam lookout position thinking that if I was painting the scene I only needed one colour – Grey. We were a grey aircraft over heaving grey seas flying low under ragged gun-metal clouds. Glancing up the cabin I noticed that the faces of the trainee navigators were also an opiate-pale shade of grey too – tinged with green.

Airsickness can be a real bugger.

The screen AEO decided with my low experience I should start lookout duties near the galley so I could serve up coffee and tea while my betters got on with the serious stuff. I think he believed I would be sick but my stomach was hardened in my boyhood simulator.

I could take a tray of coffee or tea from the galley, brace against the turbulence and high step the main spar without spilling a drop; I passed drinks to the crew, held a minute's aimless conversation before returning and still found time to pass sympathetic comments to the guys heaving up in the Elsan toilet.

Surprised by an early call to the radar tent I hurried forward; the previous occupant was hurrying aft to the toilet queue so there was no handover brief but the Screen AEO gave brilliant instruction on hunting subs in heavy seas.

Sitting sideways biting into a greasy bacon sandwich in time with a lurching fuselage was novel; we had air vents too which nicely took the edge off the previous occupant's distress. Loved it.

With guys temporarily disinclined to man the galley, being a cook on top of everything else was a challenge. Emptying lots of tinned stuff into a huge stew pot I rooted through various drawers and found a box full of spice and curry powders…

I served it in cups with a spoon because it wouldn't stay on a plate but some folks seemed to like it; although to be honest I was disappointed not being asked to cook again for the remainder of the course.

Squadron Leader Dicky Wray was our trainee captain and I was impressed with the way he got to grips with the Shack and kept ahead of the game. With a wingspan of 120 feet and an attack height of 100 feet, it meant limiting the bank to avoid digging in a wingtip but still fly a split-ass turn to attack before the target dived.

Bombing assessment was made from the tail, two bombs dropped at a 1-second interval simulated a 100-yard stick of 250lb depth charges. A '50/50 zero line was a good score because you straddled the target.

This was fine in the day but at night it was a different matter. Back in the beam, a battery of 1.75-inch flares gave light in the last mile of attack. There were 24 flares and we carried enough for about 20 runs. The empty brass casings were hot and stunk of cordite, they looked like Bofors shells and littered the floor like a brass carpet so on a reload you had to be careful with your footing with the aircraft bucketing around.

It wasn't all fun and games, on the 4th of November a Shackleton from 205 Squadron based at Changi had ditched in the Indian Ocean with the loss of 8 crew killed with 2 survivors. Prop over-speed with fire. I didn't know any of them but that night in the bar it was obvious that many guys did.

Aircrews live with the dangers of flying both in peace and in conflicts but it is still a shock when an aircraft goes down for whatever reason. The bar is very much the beating pulse of a mess and I noticed the mood held an air of determination to press on tinged with a collective sadness with no place at all for any mock bravado. No time for dwelling on it either, we had more training but sadly, we were diverted fifteen days later for search and rescue. Another ditching. Off Lands End by a Kinloss Shackleton. The Navy at the scene reported 2 survivors and on landing, we discovered there were 9 killed. Icing was a real problem too, we had no de-icing system so you couldn't get rid of the stuff, anti-icing didn't always work very well and the next month on December 21st a Kinloss Shack crashed in Inverness with no survivors of the 13 onboard. Tail-plane iced up and it went into an unrecoverable dive. The Kinloss wing lost two Shacks in a month with heavy loss of life and not surprisingly, I began to accept accidents as a normal way of squadron life.

I wasn't fussed about which squadron I joined, providing it wasn't Bally Kelly in Ireland where the weather was awful.

But first, a final check-ride. The briefing started with the MAI saying that nobody expected perfection; which was fine by me because I discovered you couldn't hide in a Shack crew and it saved a lot of trouble if you owned up to a cockup straight away.

It all ended up well with a call of a submarine surfacing flying a white flag and the MAI shouting we must break off the attack. We passed our check when Skipper Dick Wray immediately bombed the hell out of it, flag and all.

Submarines cannot surrender under the Convention!

AEOps manned the guns, with no turret on the T4 we fired the cannons on a ground range learning about stoppages. A breech stoppage tool is inserted before extraction to stop the next round feeding on top of the miss-fire. Folklore has an over-enthusiastic Siggie not bothering with the drill and cheerfully recocked and fired. Exactly as warned a fresh round fed onto the miss-fire, bent like a banana and exploded in a live belt blowing a hole in the turret. What added to his already considerable woes was a shell case hurtling backwards blowing off a chunk of the engineer's panel. A miffed Flight Engineer calmly informed the Skipper he was going off intercom, went forward and beat him up. Whether there was any truth in the story, I don't know but I could well believe it with some of the characters I met.

chapter Nine

204 Squadron v Soviets

Surfaced Soviet boat

EXACT POSITION Ballykelly Sergeants Mess bar with a welcome drink in hand.

The weather at Balleykelly gets rough. It's one of our first airfields to feel the force of depressions trundling in over the Atlantic, when this happened it rained heavily: visibility was poor and the clouds were so low you had to dig a hole to find the cloud base.

Limp garden roses bowed heads in a torrential downpour while I joined the Guinness-fuelled banter in the bar. Flight Sergeant Mike Sutton, 'Sutch' to his friends, a pleasant guy well-spoken with a ready wit told me the rain made for superb local trout fishing and I wasn't to complain. He was right. The fishing was brilliant we became friends and my BK time stretched from carefree to graphic face-time with the cold war enemy. Flying patrols from deep Atlantic to Norwegian Arctic seas, and plotting Soviet warships was a steep learning curve.

Soviets relied heavily on 'Elint' (Electronic Intelligence) trawlers which were a pain but easy to spot. Bristling with aerials they eavesdropped on NATO Coastal Bases and frequently picked up our sonobuoys. It wasn't unheard of for a disgruntled skipper to order the lead-weighted wireless trailing aerial wound out for a bit of fun and head towards a spy trawler causing pandemonium! Flying was interesting, often a piece of cake and sometimes borderline twitchy in poor weather with aircraft streaming in from missions at the same height. I had huge praise for our skippers who put up with all kinds of tense situations with bags of humour. Plotting the extent of the ice in the north Norwegian Sea was appealing. If the weather was good, sunlight sparkled off sheets of ice and large drifting icebergs in endless visibility. It was important intelligence because the southwards extent of the ice sparked off a cycle of events with the freeing up of Soviet northern ports. Their fleet breaking out is a big operation and one they liked to keep secret as long as possible.

To NATO it was surface raiders on the loose. To me, it meant flying a long way north to shadow our Russian friends in places they preferred us not to be.

Warm vests and arctic parkas were in order and landing on snow-packed runways at Norwegian bases like Bodo in the Arctic Circle was routine even in driving snow.

The Russians got cocky at times, acting aggressively towards us near their big Sverdlovs and heavy frigates.

On one sortie we headed north over the Lofoten Islands to the North Cape area flying low under a ragged cloud base with poor visibility. I manned the beam camera when a shouted warning said we were closing with a heavy warship and escorts. Our ECM recorded activity very late so we surprised them and he promptly fired off a salvo of anti-sub rockets which passed close to us. Minutes later a Russian Bison aircraft descended out of the murk passing a hundred yards away. I watched his nose jerk up at being so low and then he was gone leaving the impression that they were bugger-all good flying low in foul weather.

The Skipper calmly ordered a cup of tea while he decided how we could stick our nose into more trouble; and without any difficulty, we found it with a sub on the surface 30 miles ahead of the task force. The captain had cocked something up and my photos were so close that I could see the pipe clenched in the captain's teeth and his surly V-sign. My overall impression was that despite a show of strength, they were not ten feet tall.

chapter Ten

Northern Patrol

turret with 20mm cannon

EXACT POSITION 70 North by 9 degrees West, closing with Jan Mayen Island.

Manning the nose turret I scanned the Norwegian Sea below. A cobalt blue sky made the waters look almost inviting – until Jan Mayen Island hove into view and we could see all the ice and snow on it it. A reminder of dangerous ice from a cold eastern drift from the Barents Sea. Setting out from Bodo airbase we had turned towards the north to patrol for Elint spy trawlers in a fishing fleet.

Older hands filled me in on Russian fishing activities, our task today was finding Elint trawlers hiding in the fishing fleet; they said even I could be useful doing that. The Russians operated on a scale at odds with our British trawlers who sailed in all weathers and then went home to sell the catch. The Russians did things differently because they had factory ships. These huge vessels with dirty decks draped with cranes and gantries sailed to a fertile stretch of Arctic waters to act as mother ships to an armada of trawlers. These unloaded their catch onto the factory ship where an army of workers with sharp knives and grubby aprons gutted and sorted it. The processed fish were then loaded onto fish transporters the size of coastal freighters that sailed home to the Motherland. As a novice kipper, I assumed the whole business must have been smelly and drove everyone to consume vast amounts of vodka.

I imagined there was a lot of fuel onboard to refuel the trawlers too. Such a large fleet attracted screeching clouds of seagulls and any trawler with no gulls and lots of aerials was manned by bored KGB chaps on more expensive vodka. After spotting all of them and photographing them shaking fists we set off for any other business to the North.

It's pleasing making a nuisance of ourselves and after serving tea all round, I bounced back to the nose just as we came upon a group of Russian Myrny class whalers. Idly fingering the gun firing stick as we closed for a visual ident, I was given a lesson in diplomacy by our Skipper. I liked whales and faced with a choice of roast beef or grilled whale I would plump for the beef any day. But not our Russian friends. After harpooning a whale they pumped it with air, stuck a flag into the twitching body and went after more. A ghastly business. The whales were still alive, fluttering their fins feebly as they slowly died.

They also looked like a submarine on the surface.

I cocked the guns which was a no-no and requested clearance to shoot one to put it out its misery and get some gunnery practice. The response from the skipper was immediate. The one-way conversation included phrases like 'It's a bloody International Incident if you put one round anywhere near one of the bloody things'... and so forth. I acknowledged with good grace, and turning around I saw the two navs and both pilots looking at me as if I was out of my mind. Taking this as a definite no, I apologised for getting a bit carried away.

The AEO thanked me for my generous offer to shoot up whales, rightly admonishing me for cocking the guns and said he liked my spirit but would I man the radar? 'Man anything except the guns' is what I think he meant. These irksome international strings attached to whaling seemed over-complicated to me but I picked up huge tips on radar operations off the Norwegian coast.

Our ASV 21 radar had a 4-degree beamwidth so it would bounce off something as small as an attack periscope to produce a large echo. The same was true of small rocks and islands off the coast which made them look huge on our screens and comparing these returns with a chart of the coast was a nightmare. But the human brain, even mine, is an amazing computer and with practice, we developed 'Norwegian eyes' where the whole thing made sense.

Our ageing aircraft provided a window on good captaincy. I noticed an immediate acceptance of a serious situation; nothing was rushed and without any apparent extra effort from the skipper, options tended to float to the surface.

I was in favour of options rather than our wreckage floating to the surface because some of our Shacks were getting long in the tooth.

Sometimes a Griffon belched out oil, smoked like hell or even caught fire. Losing an engine was no drama and I recall skippers shutting engines down and asking for a cup of tea in the same tone of voice. Many incidents were never recorded in our logbooks because attention-grabbing situations were forgotten after a winding down drink and we filled our logs in at the end of the month. One incident comes to mind because of the sang-froid of the skipper. It started way off the Norwegian coast well into our search area. The starboard outboard engine started making loud bangs and belched oil and sparks. The Skipper shut it down but black smoke still trailed past the tail. Although heavy with fuel and a full load of weapons and sonar buoys it was no real worry, but, with an engine still smoking we turned back towards the coast and had a brew. A good decision by the Skipper because the inboard engine on the same side decided to belch and trail thick oily black smoke as well which meant we had a problem. Throttling back reduced the smoke. The skipper ordered a mayday call on HF to help things along and then the fun started.

The Flight Engineer sorted everything out and I must say he didn't bat an eyelid shutting down systems and shunting fuel all over the place. But he had survived the last war despite ending up in a Stalag so I daresay he found it all a piece of cake.

The Navs were brilliant, one quickly plotted course south-east for the nearest airfield, while the other set up a safe jettison and as we turned on a heading for the coast I headed for the galley to make tea for everyone. But never made it. Our AEO shoved me on the radar pointing out I had been in Norway for weeks and knew the coast. I knew what was coming. I grabbed a chart and the AEO jabbed a digit on our estimated position while I scrutinised every bit of sea ahead of his hairy thumb.

I heard the nav say, 'Captain safety height at the coast is 4000 feet and we're at 400 '.

'Right! better sort something out then.'

'Radar check ahead please.'

'All clear no contacts,' from me,

'Super, Jettison safe please,' and away went a lot of expensive gear. It was like an ancient sacrifice to a Nordic God made by a bunch of unwashed Vikings asking for a bit of slack in a tight spot. Anyway, it helped. For a while.

The skipper advised below 200 feet that we were in a spot of bother committed to flying along the coast and turning up a fiord to stalk an airfield. Or ditch.

'Start throwing everything out we don't need' was a beaut order.

The electronics gear on a long shelf above our stations felt the fire axes first. I heard them cutting through cables followed by heavy footsteps past my tent as Siggies staggered aft to chuck stuff out of the beam windows.

We could see the waves, we had maps, radar and VHF, so no surprise when the fire axes started on expensive nav gear. Throwing stuff out became an industry, someone in the beam tossed flares and flame floats out the beam window making a brew between bouts of hurling. A stretch of sea free of obstacles ahead was a good time for a pee but I got caught in a conversation about dinghies (we always carried one on the top bunk) as I was squeezing between two guys undoing it to get to the Elsen toilet – which I hoped was still with us.

'Give a hand to chuck this out Wardy.'

'But we might be ditching.'

'No prob's there's one in the wing on this kite, should come out automatically when we ditch.'

I didn't like the '*should*' bit, but I liked the idea of throwing expensive stuff away. It needed a lot of shoving and we had to move the Flight Engineer aside who was idly tossing baked bean cans and tins of Danish Ham out the beam window.

Back on watch, I was squinting at the screen when I felt a hand ruffle my hair. I swivelled around to see our smiling Flight Engineer with a hand behind his back. I told him to buggar off and turned just as he smashed an axe into the radar transmitter above me.

Miffed at my radar being smashed without being asked, I grabbed his neck and the axe, threatened to brain him with it and shooed him forward to his panel. I was learning that anything can happen in Coastal, but not worth complaining about even if it was a pain in the ass when you are about to ditch. Closing towards the inlet leading to Orland field and leaving on the seabed a trail of assorted torpedoes, expensive electronics kit, tins of baked beans and Danish ham, we settled comfortably back into our crash positions for the third time. Later in the bar, we got legless and in the morning were informed there was quite a bit of repairing to do... My logbook entry: 'Take off Bodo Land Orland'. Times ...

chapter Eleven

Capricious fortune

WITH EXPERIENCE, I noticed Some Captains could grasp a situation by the balls in a seemingly casual manner which was demonstrated on a Search and Rescue (SAR) sortie one Christmas. Being single I volunteered for standby and on the 27th of December, we scrambled to look for three missing guys in a boat off the south coast of England. I manned the wireless and promptly got on the morse key to Rescue Co-ordination Centre (RCC) boys for a sitrep. It didn't look bad but the picture slowly changed. In voice contact with the RCC, I heard a lifeboat reporting. The target was a dark blue small boat open to the elements and three people had gone fishing early on Boxing Day; they were expected home by mid-day. It was now approaching dawn on the 28th and they couldn't be found by lifeboats or a helicopter search. Sea fog further out had cleared. Some pretty cold guys out there.

The Skipper and the two Navs went to work and I kept the channel open requesting past weather. The Navs went to conference intercom and I earwigged to butt in with info. In the beam, with flares loaded and flame floats topped up everything was ready so tea all round.

I always held our Navs in high regard but these surpassed themselves. One set up a plot and the other searched almanacks, working out tides and current to get some idea of possible drift. The start point of a creeping line ahead search was now critical: the lifeboats had searched on local knowledge which we didn't have.

Dawn was a sliver on the eastern horizon when radar plotted everything afloat and the Navs got a very accurate fix on coasting out to steer us to their calculated start. The facts were not good. The lifeboat had gone home at dusk, a helicopter was on standby and we were the last hope.

It was needle and haystack stuff with a visual check of all contacts. The Navs came up with the goods with past weather. Updating assessments of tides and winds, they switched the creeping line ahead and the radar came up with a faint return. The Skipper had taken it all in and the Lindholme Gear was ready in the bomb bay. It would be dropped at 140 knots from a height of 140 feet and the survivors would drift onto a rope joining the two containers with a dinghy. All of us knew how difficult it was boarding a multi-seat dinghy and the Skipper was highly aware of it. The Navs were bang on. It was the target.

We roared over the little boat dropping flame floats and I was ordered to the tail position with the explanation I had young eyes. Three listless guys were huddled together, one gave a small wave. But I had seen hypothermia a few times and I didn't like what I could see and asked for another on top and got it.

My feelings were echoed by the nose lookout and the two pilots and this time I had the binoculars on the lad in the middle of the group. Not good. Classic listlessness – unaware.

The Skipper made an immediate ballsy decision. No Lindholme drop – too risky but we had located them. Someone else had to rescue them.

I forget the name of the small town we be*at* up to say good morning to but remember it was early and empty. There was a jetty with boats moored alongside and we flew low over the rooftops with the Skipper altering the power and generally enjoying himself rattling the tiles. After the second run, passing overhead the jetty we flew out to sea in the direction of the survivors dropping flame floats to mark the way and firing off red Very's. The locals, quick on the uptake sprinted towards their boats; a pale speed boat quickest off the mark left a long wake past our markers.

After dropping more floats and a wave to the chaps in the blue boat we set off for BK feeling chuffed they were OK. The Skipper and Navs had been brilliant in my opinion.

My life changed in March when Squadron Leader Clive Haggett who I liked enormously said he wanted me to join his crew. I knew them all and in particular, Nat Duffy, a Siggie and the workup with them all was going great guns. I was due to fly early on the 19th of April but on the afternoon of the 18th, Clive phoned me to ask a favour by flying with a 210 Squadron Skipper. They were a Sig down and due to fly to Norway very early in the morning. It was disappointing but couldn't refuse and besides, I wanted a Norwegian fishing rod which he knew. After some shut-eye, I left at an ungodly hour arriving in the dark at 210 Squadron to meet the Skipper and crew.

Scrawling my signature without bothering to print my name, we flew to Norway. Hours later, my normal 204 crew took off to play with a submarine. But they never came back. They crashed into the Mull of Kintyre and were all killed. When we landed in Norway the Skipper came straight to me with a signal in his hand and broke the news. I felt gutted, the Norwegian trip was cancelled and we flew back first thing in the morning.

I had a bad night. But it was nothing like the feeling the next afternoon when we flew over 204 car park and I saw the parked cars of the crew.

A final confirmation.

All the squadron office dwellers had gone home and nobody was around when I ambled in to drop my gear. Then it dawned on me that nobody had seen me fly off with a different squadron crew in the early hours.

It was a custom in the Mess to put on a few barrels at a time like this and charge it to a single chap who wasn't around anymore and then write it off. Word had spread that I was killed so Sutch and friends were stuck into a whole lot of drinks on me when I strolled casually into the bar. I well remember the stunned silence, the choking feeling as a great cheer went up and everyone threw beer at me: I couldn't remember anything of the rest of the night.

It was a jagged moment for the squadron. Boss Kent was magnificent, displaying true leadership but the sad truth was that with so many dead we had to choose which funeral we wanted to attend. When Nat's name came up I volunteered immediately; he was buried in Southern Ireland and I said goodbye to him alongside his family and many others.

chapter Twelve

Pleasant Endings

THE LETTER in my hand ordered me to report to the AOC for an interview for pilot training. This was years early and I read the letter five times just to believe it before bounding in to see the AEO leader. He remarked they must be bloody short of pilots, advised me to get a haircut and said I was to join Gordon Smith's crew. Smith was an ex-Signaller with a reputation of being a good Skipper but hard on newcomers. Upfront was co-pilot Bill Mott and two Navs Peter Marks and Pete Curry who I became good chums with.

The AOC a gentleman, put me at ease by asking how things were at 204 after our loss and was I enjoying Ireland? After comparing brown trout fishing he invited me to air my views on future nuclear-powered sub-detection. He was in luck. I had several. I held the opposite view to those who thought fast nuclear boats were undetectable, only recently did I raise a few laughs over one Guinness-fuelled theory in the bar.

My theory was loosely based on my African experience. In Nairobi, I met a gin-soaked old hunter in the Long Bar who claimed he could put his ear to the ground and tell what sized elephant was moving in the bush. Where it was precisely tended to be something of a guess. But, he was convinced if his gun bearer had stayed put with his ear to the ground he could have had more success: although he conceded you needed more than two heads on the ground to fix a fast-moving bull elephant. His theory was a little flawed but the principle was sound. I wasn't foolish enough to mention this though.

It was a question of wide-area triangulation and acoustic spectrum. I ventured we needed to develop huge passive sonars that worked on a variety of frequencies like reactor cooling pumps and irregularities in shaft rotation as opposed to cavitation alone.

We could air monitor these sensors on the sea bed at strategic locations.

If the AOC thought my enthusiasm lay ahead of my knowledge he didn't show it; he smiled and wished me luck on pilot training. Couldn't believe it!

Sutch and a host of friends couldn't believe it either. They dragged me off to celebrate and Pete Curry who had been invaluable with advice gave me the biggest cheer. But, no time to savour anything, Gordon was chosen for 'Select' crew status. It's given to less than 10 per cent of crews following a gruelling check and real swagger stuff for a captain. Preparation was rigorous. Day attacks were easy compared to night ones and we had to be good at everything. Stripped to the waist sweating under a dim red light manning flares was arduous. After firing, hot empty shell cases leaking cordite fumes were levered out onto the floor and another Siggie rammed in fresh ones. It's like being a gunner in Nelson's navy: air thick with cordite, sweat dripping down you in tiny rivulets through clinging dust. Bandanas tied around the head kept the eyes clear but only just. I held a hand on the hot breeches counting the recoil, a gap in the thudding was a miss-fire and on reload it needed turning through 90 degrees. If no luck on the next salvo you had a problem. The unexploded flare was removed, taped over and gingerly sent down the flame float chute. With brass cases littering the floor it wasn't easy but flares were a picnic compared to photo flashes. Fired from a discharger in the roof they were lethal in a miss-fire. We had one and Graham Seaman an experienced Sig calmly held a roll of tape while I eased the photoflash out, trying not to show the whites of my eyes. Keeping a good foothold amongst the casings on the floor I waited for him to tape the end before it went down the flare chute. Despite taping the end, the slipstream dislodged it and it exploded with an almighty flash just feet below us, blackening the fuselage. If that had gone off inside it would have been a disaster.

We were awarded Select crew status and I bounced off to Biggin Hill. Interviewed by a friendly selection chap I was surprised how interested he was in my Coastal time. Had it helped? He confirmed I would be commissioned, thanked me for making the best of holding and said my enthusiasm had been noted.

The RAF had more than kept its word, barely sixteen months on a front line squadron had passed. Even better, Gordon let me fly the aircraft for hours, and the entire crew expressed pleasure at seeing me in the co-pilot's seat instead of the galley – the claim my stew caused more airsickness than the entire North Atlantic weather system I pretended not to hear.

My short time in Coastal Command was over and despite taking friends from me I had an affection for the Shackleton, a truly gritty aeroplane. My affection was not only for the aeroplane but also for the crews and their comradeship. The fallout during my short operational period with the Kipper Fleet was alarming, sometimes I doubted making it through. I salute the forty-one brave RAF and Commonwealth men who were lost during my time.

Smoking Red – Part Two

chapter Thirteen

RAF Primary Flying

EXACT POSITION RAF Church Fenton Officers Mess Bar

The atmosphere was light-hearted bordering on rowdy. It was early January '70 and I was now a Flying Officer because the RAF took into account my enthusiastic holding time in Coastal Command and despite the fact I mostly brewed the tea on Shackleton's, the RAF kindly backdated my commission.

In the bar were the pilots of 255 Pilot Course: Bill Burborough, John Bartholomew, Phil Flint, Chris King, Steve Jarmain, Ian Davidson, Steve Riley and Nigel Voute.

And all of them were better at maths than me. That was clear in the first hour on the first day of our ground school and I wasn't smarting a bit.

My poor show on these 20-minute morning speedy-calculation tests was the unfair distraction of a line of shiny Chipmunk Trainers drawn up outside. Every few minutes my gaze lifted from boring calculations to dwell on their silver shapes.

Steve was the youngest trainee pilot and judging by the youthful expression on his face, should still be at school. He was hoping to pass his car driving test soon. Bill was a little older and I slotted somewhere in between. A happy mix of direct entry guys with a sprinkling of ex-Airmen, Apprentices and a Boy Entrant. We survived Officer Training in 1969 followed by aero-medical runs in decompression tanks exploring lethal hypoxia observing each other making asses of ourselves.

The question on everyone's mind was what were the Instructors like? On cue, a few drifted in to have a beer with us. I was on the fringes of a group trying to look keener than everyone else when a beer tap exploded. A tall chap in a brown Harris Tweed suit with a shock of unruly blonde hair had sneaked in and taped a black powder charge to the pump. He stood grinning at the bar and introduced himself as Dai Heather-Hayes. Instructor.

As fledgling pilots and navigators, we received daily doses of meteorology, navigation and aerodynamics. I had boundless enthusiasm for met' having flown in atrocious weather in Coastal, which tilted the academic balance because I was barbaric at maths only scraping through with few marks to spare. Each day started with a wake-up shake and a cup of strong tea from our batman. Breakfast, then off to Ops for 'morning prayers' – the weather forecast for the day – followed by an outline of the flying programme.

Assembled in the crew room in squeaky new flying kit we waited for our instructors wondering who would get the hard guy. Bound to be one.

Eight Instructors collected students with varying degrees of enthusiasm and I was left wondering if someone had re-marked my maths paper with a lower score.

Maybe the Education Officer was on his way right now to make me do a re-sit?

I idly wandered over to a window to watch engines splutter into life.

Boom! The door crashed open. Framed in the doorway was the tall figure of Flying Officer Dai Heather-Hayes; he breezed into the room with blonde hair askew still dressed in his brown tweed suit complete with gold watch chain.

'Right Ward do you think you could get into the Queen's aircraft without wrecking it?'

My RAF flying training had begun.

Dai went through the start drills, a cartridge banged in the starter the engine fired, after-start checks were simple and we were good to go in minutes. Lesson one was familiarity with engine power, attitude in level flight and how to trim and fly steep turns. After ten minutes Dai got bored and threw in some aerobatics to liven things up and after a few loops and barrel rolls asked me to try trimming. I was just getting the power and trim sorted when I heard a joyful 'I have control' from the back, the stick whipped over and suddenly we were in the middle of a dog fight with two Naval Chipmunks returning home.

Dai plunged into them and I got to know all about turn radius because we ended up on somebody's tail with the scrap ending in a prolonged tail chase with us in the middle. Marvellous!. In just twenty minutes I was completely bowled over with RAF training and after Dai demonstrated a circuit I even managed a few without disgracing myself. But after grabbing a coffee the de-brief didn't go as expected; when Dai asked me what I had learned I started droning on about power and trim and he stopped me by belting me with a rolled-up map.

'No Ward' he said with a grin. 'Never take on a pair unless you have the sun,' and proceeded with good advice on 'bouncing' other aircraft.

And so ended my first lesson of aerobatics, dogfighting and somewhere in there a bit of straight and level. After coffee Dai enquired whether I could start the Queen's aircraft without blowing it to bits...

RAF Rufforth to our north was used as a relief landing ground so we didn't clutter up the circuit at Church Fenton. The Primary Flying phase is short but marvellous fun with days of stalling spinning and aerobatics followed by a quick bath and a stroll down to the bar to laugh like hell at our mistakes. After six flying days and many dogfights, Dai ordered me to Rufforth for circuit bashing and after a couple of landings, jumped out secured his seat and told me to buzz off round the circuit while he had a cup of tea.

High-fiving each other we all soloed, I flew with Dai H-Hayes for all my dual time; he was enormous fun to fly with and his excellent instruction boosted my confidence. The only problem I inherited was assuming that bouncing aircraft and impromptu aerobatics were normal. But Dai was far too astute to let me get cocky and if I pushed my luck too far he would hold the throttle closed at some inconvenient time to make me go through forced landing drills.

I enjoyed RAF Primary Flying. Having an instructor with an immense character was a bonus and mess evenings were fun with Dai firing his black powder shotgun into the ceiling and blowing up beer taps. Even cricket took on a new dimension with him. One dining-in-night with a broom handle for a bat and his chum bowling half-pint glasses down the table he showered the place with broken glass – the rumour of him flying past the CO's widow standing in the rear cockpit giving a salute with the pilot in the front crouched low has real substance.

You would have to search hard to find more mediocre pilots than us but none cared. We were off to the picturesque Vale of York, known in flying circles as 'the Vale of Death'. The RAF liked to keep trouble in one spot and with flying schools at Leeming and Linton-on-Ouse crammed with Jet Provosts – avoiding the place was a good idea.

chapter Fourteen

RAF Basic Flying – Linton-on-Ouse

WE ARRIVED for jet training just as a T4 Jet Provost (JP4) suffered engine failure at RAF Leeming to the north. The pilot ejected safely opening the crash statistics in the Vale for the year. On our first morning, the Chief Instructor told us to be careful with throttle handling and not screw up like the Leeming jet yesterday—but he spoke too soon. Squadron Leader Dawson and Flight Lieutenant Ivor Gibbs, responsible for our training, were bidding us a welcome when the crash alarm sounded. A Linton student crashed his JP in the undershoot which put the 1970 crash score to evens with Leeming on day two.

Ivor Gibbs was one of the best guys you could meet in aviation. He was ex-Coastal for starters and interested in my African time. Would I lead an expedition from Linton to climb Mt Kilimanjaro? He outlined the course: The Queen paid us and for our fuel, it was the best deal on the planet with basic handling in aerobatics, pilot navigation, Instrument flying, and formation flying with the weather dictating which bit we did. Each of us would be allocated to a Qualified Flying Instructor (QFI).

Ian Mackenzie was a tall good looking chap with a lantern jaw and of stoic disposition to have read my training file and still want to fly with me. He was the kind of guy you used to make films of heroic Biggles and not saddle with an incompetent.

To fly jets we needed ejection garters and better helmets, the garters were worn below the knees and were sturdy affairs with metal D-rings attached through which we threaded two frangible lines anchored to the floor. When ejecting upwards these lines drew your legs together so they didn't flail about in the slipstream and mash up the hip joints and leg.

Ground school was over in a flash – the JP is a simple aircraft. Brian Penton-Voak a Navigator taught navigation underlining because we were pilots we had to keep things simple. For some reason, the Boss nominated me as the course leader – responsible for liaison between us and the training staff. Looking after everyone's interests would fall to me. I found the idea spectacularly unappealing because I probably needed looking after more than anyone; a sentiment shared by the entire course but they still wanted me to take it up which I found amazingly misguided.

On April 1st we were welcomed chat by the Chief Flying Instructor Wing Commander Clark and then Ian briefed me. We would use JP3s with unpressurized cockpits which meant you had to watch your intake the night before. When the cockpit altitude climbed: you passed wind a lot.

Compared with the Army and Navy, the RAF had much experience with farting and it was never the social problem it could have been even though we sat close to each other. Wearing oxygen masks all the time helped in severe situations.

Much thought went into the design of the JP. New pilots needed to experience the handling of operational jets but still retain a lower speed for stalling and approach so we didn't crash too often.

The smell and sounds of a jet flight line are different from a piston line and the first time a trainee pilot walks with his instructor towards a line of jets his senses are more vibrant and alert than ever. Burned kerosene wafts by on a stream of hot air, a sudden crump of fuel igniting on a starting engine and the rising crescendo of a jet moving off the chocks all register in the mind and are never forgotten.

Anyway, it was April Fools Day, the weather was good for my first flight. Ian showed the walk-round checks and clambering in I was surprised how close I was to him. Starting a jet was easier than pistons so we were ready in no time and taxying was easier with a nose-wheel – with no engine in front we had a better view.

It was a case of 'doing and feeling'. The RAF didn't believe in molly-coddling with endless theory, certainly not on Basic Flying because this is where you were judged suitable to wear the RAF pilot's brevet. You pulled and pushed shiny levers, you knew what the interesting buttons did when you gave them a poke and you took the aircraft to the limits – or you took up wall-papering instead.

The acceleration's slow but the speed kept building. Ian was a patient accurate flyer and when he handed over control my efforts felt slovenly. But Hey! I only had 10 minutes of jet time and a whole year to improve. I was foolish enough to mention this and Ian promptly agreed I would need all of it. The other unfortunates who flew with me were from mixed flying backgrounds.

Elwyn Bell a South African was a carbon copy of Ivor Gibbs. Stefan Karwowski was the son of a Polish pilot who fought with the RAF. Al Curtiss was ex-V-Force with bags of experience, as was Noel Parker-Ashley. Pete Frieze was an ex-Lightening pilot with a light touch and a quick brain. Nobby Grey ex V-force had a relaxed but discernible air of determination and it didn't surprise me to hear years later that he put in a courageous performance flying against insurgents in the Omani War. Dudley Carvell, a young astonishingly calm and friendly instructor I learned was a 'Creamy': these were exceptional students who after advanced training were returned into the system as instructors.

All good guys. Especially Ian, who undid my bad habits and had me soloing in quick time.

After the first solo, we flew 'sector recce' like early days in the RFC where new pilots familiarised themselves with landmarks in their sector.

Training can be dangerous. It sank in when a student from RAF Cranwell crashed a JP4 on a night approach and was killed; a Gnat crashed at Valley too in January so every pilot training school had prangs. But seeing as we were in the business of throwing aircraft around the skies nobody gave it much thought. For our safety, we did spinning and aerobatic checks before soloing which I looked forward to and with all these checks I couldn't see much going wrong.

Early for Ian's brief on spinning, I sat drinking coffee in the briefing room gazing at his amazingly detailed diagrams. All pilots know about spinning these days but I knew from old dog-eared books this had not always been the case. Stalls and spins had dogged aviation from the beginning, a case of flying into the unknown if you got into one but the sobering fact was that

Orville and Wilbur knew less about aerodynamics than I did. It's remarkable how they put it all together but it was as late as 1918 when the spin recovery technique was fully understood.

A bit late for the brave guys on the Western Front – a spin was not a good thing to be in when Von Richthofen pitched up. The Red Baron probably didn't know much about the aerodynamics of a spin either but he knew a damned good target when he saw one. However, after 1920 much more information was available because pilots carried parachutes and floated over the wreckage nursing a strong desire to talk about what went wrong.

Ian breezed in and explained how to get into a spin and how to get out of it. For some reason, it drew out a fascination and strong desire to know everything about the whole business. Suitably fired up we took off into the Vale mindful of the fact that eight others on the course were doing the same thing.

Keeping a sharp lookout I watched fascinated as a toy-like JP tumbled down close to us: eventually, Ian judged it safe. Speed back. Big judder, a full boot of rudder and off we went. With the nose about 45 degrees below the horizon, we were descending at about 800 feet a revolution and I was totally into spinning. Great stuff.

Time to recover. Throttle closed. Centralise controls. Ailerons have to be centred otherwise the spin speeded up. A big boot of rudder, stick forward and out we came.

Ian chatted over problems of miss-control which helped enormously when I got myself into trouble later. After practising umpteen solo spins I was surprised when a few guys admitted over a beer they just flew aerobatics and didn't spin solo. Too risky. Why foul-up? Astonishing but I suppose different characters were emerging. We were probably into a subconscious divide between those more gung-ho who hoped to fly fighters and those who preferred something bigger that didn't spin or turn upside down. I noticed too that as the standards became more exacting, moods changed.

During training, it's rare for everything to go smoothly and everyone at some stage has a hiccup or fails to achieve the lesson aim without a hitch. There's no magic way of rectifying these situations other than being honest and giving friendly support, although I did notice a sense of humour helped. I needed a dose myself after I hit a huge Fijian on the rugby pitch so hard I broke my collarbone and it's difficult to fly with an arm in a sling. When my collarbone was barely healed, I convinced the Station Doctor I was fine and with a much padded flying suit, I started flying again. I was relieved to get airborne because Basic Handling Tests (BHTs) loomed and they were the first major important hurdle to clear in basic training. This was where guys fell by the wayside but remarkably the whole course made it over this high fence.

chapter Fifteen

Trouble with Enthusiasm

OUR NEW Jet Provost Mark 5s known as JP5s was more streamlined than the JP3 and 4 and being pressurised led to less breaking of wind. Stef Karwowski was giving Ian a well-earned break from me and we revisited spinning because the JP5 was different to handle. It wasn't passed for spinning initially because of large oscillations when the outside wingtip and canopy became partially un-stalled at the end of each turn. This meant a lot of pitching and rolling moments, not at all fit for ham-fisted students like me.

Fitting nose strakes increased the yaw rate and angle of attack so everything stalled nicely. Carmex's non-slip walkway paint – as rough as a bear's ass, slapped on the outer wing leading edge solved the wingtip un-stall problem by stalling them earlier.

Stef emphasised during recovery the stick should be pushed forward only to the mid position – unlike the JP3 otherwise, we would end up in an inverted spin which was a definite no-no. Not to be attempted. Well not deliberately anyway.

It was great fun. The juddering yawing moment and total abuse of the aircraft was intoxicating stuff and I got carried away and had to be stopped from endless spins. Instead, he briefed me for a solo flight to do it all over again.

No sweat. Nothing ever goes wrong on a solo repeat – usually.

Everything went swimmingly until the first thing I did was the last thing I was told not to do.

I had to. I mean who wouldn't be curious?

Slow speed, bags of nose up, a boot full of the rudder and away we went. A piece of cake. Up until I deliberately shoved the stick forward to the stops to see what would happen (moments like this hold a curious thrill of expectation until the consequences arrive)

Stef was bang-on. After a touch of shuddering which I didn't like, I flipped onto my back and straight into a 'no-no' inverted spin; I wished I hadn't done it but too late now. Problems mounted quickly because I was a bit disoriented and my shoulder harness was nowhere near tight enough. For a brand new aircraft, an amazing amount of dust flew around the cockpit and suspended in my straps upside down spinning merrily away it wasn't a good feeling. My feet flew off the rudders and I had difficulty keeping my hand on the stick because I was higher up the seat than normal thanks to the slack harness.

A part of my mind – except the huge bit in a straightforward panic – grasped the nightmare possibility of explaining to the Chief Instructor why I ejected from one of his perfectly sound JP5s.

Keeping a good lookout was a problem up-side-down because hurtling earthwards I had to look up to see what was below me. Like it or not regaining control was now the deciding factor between wall-papering for a career or flying.

Shoving my feet against the rudders under negative-g was a new experience but I managed it and eventually got a grip on the stick with my fingertips and pulled it back to the centre. And Hey! After a brief juddering roll the horizon ended up where it belonged and I was out of it.

If you get bucked off a horse providing you haven't broken anything much you get back on straight away to keep your nerve. I extended this principle to my present situation. Had to do it. This time I was in the zone with tightened straps and my feet hooked into the rudder pedals. I spun properly. The right way up. Everything was going well after five or six revolutions and compared to my previous experience it was almost boring – until I decided to try full in-spin aileron – rapidly – just to see what would happen.

Boy did it tighten up. I may have been slightly overconfident because I burst into a cloud that looked well below me when I started but I was hugely enjoying the fast spin rate and probably hung on too long – that's the trouble with enthusiasm. The high rate of rotation on instruments with no friendly horizon was a bit unnerving to be perfectly honest but despite a sinking feeling in the stomach I had a stab at recovery with much confidence and was mildly surprised when I burst out of the cloud the right way up.

The third spin was perfectly normal and an anti-climax because I did what I was supposed to do.

Had to be honest when Stef asked me how it went. After all, he had authorised me to go off solo. But we were pilots, not priests, so seeking moral high ground by telling the truth didn't enter into it. No. It was more of an insurance policy in case someone had seen me tumbling upside down in the Vale and took my number. I managed a cheesy grin, saying I did get the stick position slightly wrong but I sorted it no problem. Fun though. With four courses of pilots running, happy hour was a noisy melee of opinions bullshit and beer. Characters like Bob Husher on the course behind and Stu Ager ahead were buoyantly good company. The simple truth was that we were not competing against each other – if we all met the standard we would all get our wings.

We entered the more advanced phase of our basic training with me owing Ian a great deal for knocking me into shape. I was to fly with Al Curtis and hoped he would still be speaking to me after a few weeks together.

Night flying was a disciplined affair and more so since the fatal night crash at Cranwell. We were encouraged to get rest during the day to be sharp at night but views differed on this. Phil decided to party most of the night and sleep in the day but overdid it, fell asleep flying in the circuit and had to be woken up by an irate instructor.

All new pilots feel an ethereal detachment on a night with lights of towns sparkling below and pricks of light from the stars above. Sometimes the controller's voice in your ears intrudes on your newly discovered little world and you almost resent them for it. But hidden dangers lurk unseen in a nimble jet; a steep turn in turbulence with patchy cloud cover can lead to disorientation. After passing our night flying tests we moved on to something I had been looking forward to – Formation flying. The stuff of dreams. Early military flyers slipped into the habit of formation flying for the simple reason that flying as a pair or more gave mutual support. In later times with a loss of radio and no navigation aids, short of fuel; it was a life-saver.

So the RAF insisted you did a lot of it and they made sure you did it properly.

Al bravely took me for the first sortie then Pete Frieze ex-Lightnings who made it look like a piece of cake and I loved it even though I started to get bunched up without realising it. When constant fiddling almost led to a collision, my instructor casually chatted to me about beer and cars. And you know what? – I relaxed completely.

Hard on the heels of formation, came low-level flying and navigation. No flashy arcs and bearings just a line on a topo map marked out in ticks of 4 miles per minute at 240 knots, or 5 miles at 300 knots. Great stuff. Weather framed our activities at low-level and rolling decision-making became harder with higher speeds which was the aim of the game. Noting escape routes in tougher terrain would come later at a much higher speed if we ever made it to the ground attack role.

We sensed the end of the course was in sight. No one had been chopped. No failures. Unusual. But the strain was on for some. Team spirit was now everything. And when Ivor took me aside one day remarking we should be looking at Kilimanjaro I took it as a hint that with no screw-ups we would all be graduating.

That night I reflected on our progress. My mind wanders to the words of Ernest K Gann in 'Fate is the Hunter' where he describes how the training changed him and his novice brother pilots. It would be fair to say that change had fallen over us all too. From an aspirational but uninspiring bunch that started, we had retained our enthusiasm but we had grown warier of the sky and its' hidden dangers. Some had been utterly shaken over a bad experience but were now stronger for it. For starters, you left Cumulonimbus clouds alone. Without weather radar, all of us had blundered into one and experienced the lightning flashes, surging up-currents and violent downdraughts that left you feeling insignificant. Relief at getting out of it and a return to normality left you shaky. Classrooms taught us icing could be rapid with a deadly legacy of increased weight and lost lift. The theory of distorted wing shape was fine but the practicalities were deadlier especially when the engine turbine heated up. Throttling back to avoid over-heating meant you lost the power to climb above it all.

But we had taken the aircraft to its limits, thrown ourselves into aerobatics and stalling, spinning was second nature and a piece of cake. We could even navigate at low level without getting lost – most of the time – and if we lost radios above the clouds with no nav-aids and were uncertain of our position, we could slip onto the wing of another aircraft without hitting it and get down in one piece.

We were getting close to being RAF pilots.

Ivor and I threw ourselves into planning our mini-trek in East Africa to climb Kilimanjaro. Once in the bush, I would take over. I suggested a plan B for Mount Kenya with its excellent relations after Uhuru (Independence). But whatever happened we would climb something high in Africa for sure. Getting the guys out training was like herding hens but we managed a few weekend treks to get used to the pace needed – sprained or broken ankles on the high mountain would be a show stopper. I was scanning a map when Stu Ager bustled in with his customary grin to run something by me. He had come top of his course, winning the Eustace Broke Loraine trophy. Winning this valued trophy meant he had won the sword of honour position – the parade commander – and a lot could go wrong if you made a balls-up.

Ivor Gibbs met me on my last training sortie informing me Mount Kenya was on. I was secretly pleased because although it was the less climbed cousin of Mount Kilimanjaro, the scenery and trek up were better according to my mountain rescue chums.

Briefing my Final Handling Check, the Chief Instructor, Wingco Clark informed me he was coming with us up Mount Kenya so I had better pass my check ride. He was looking forward to it. 'No pressure' he chuckled.

It was a good omen because it went well. He shook hands. I had done it! My journey from horseback to a cockpit was almost complete. I was to gain my RAF wings. I went straight into planning the routing for Mt Kenya, turning in an awful performance during our aerobatic competition. Euphoric at returning to Africa, I fell out of two reckless stall turns and flew a loop below authorised height. Got told off. Was happy beyond measure.

The biggest reward was the pilot's brevet but the RAF gave many prizes to keep an edge on the competition and the CI bustled in to announce them.

Steve Riley, sharp of brain and the last of us to leave school by a considerable margin won the Ground School Playfair Trophy. I called him an inky-fingered swot.

John Bartholomew was the aerobatic competition winner. Despite running into severe headwinds during the last months of the course, he had stuck at it. We had chatted and I felt he was gutsy and told him so. God bless him, he starred – a popular winner of the RN Trophy for Aerobatics. Steve Jarmain, our most accurate flyer won the Fuller Trophy and Six Course Trophy for best in general instrument flying.

Now for the big one. The winner of the trophy for Best Pilot. The Eustace Broke Loraine Memorial Trophy went to the student with the highest standards in flying, ground school, games and just about everything else. It was also for meritorious officer qualities – so I wasn't paying attention. Some of our guys were really good pilots and I knew they had done well but I couldn't help feeling whoever won it, wasn't counting on the sword of honour bit of taking the whole parade that went with it.

I had been course leader organising everyone for long enough and felt smugly complacent about taking my place in a line to get my wings. Risking a balls-up by marching a whole column of men straight through the Station Pipe Band was for someone with a career.

chapter Sixteen

Wings Parade

EXACT POSITION standing on chalk cross marking Parade Commanders spot

The familiar weight of the sword felt comfortable in my gloved hand. Behind me the Officers and men of Number 1 Flying School stood to attention ready for the VIP.

Bill Burborough Parade Adjutant stood straight-backed and proud flanked by Phil Flint, officer commanding number One Flight – our boys. Behind him, Bob Husher and Malcolm Young commanded the other two flights. I felt immensely proud, not for myself but everyone on our course, we had all gained our wings. We had helped each other. There had been no fatalities, no failures – a rarity.

So far I had not messed up. The RAF Regiment band looked magnificent. A few practices had helped and they no longer feared me losing the plot and marching a host straight into their shiny ranks.

On cue, Major General Bell, Commander Third Air Force USAF in Europe Supreme Allied Command, drew up in his limo and marched to the podium looking every inch a leader. I took a deep breath, marched the entire parade forward to salute him and asked him to do the honour of inspecting. His easy smile relaxed me and everything went swimmingly; the rest of the day was a blur talking to our families and Mother and Father were the proudest in the land that day.

We would split to Advanced Flying Schools. Some to RAF Valley in Anglesey to fly Folland Gnats on the path to fighters and some to RAF Oakington in Cambridgeshire to fly the heavier Varsity, then on to Hercules or the V-force. Steve Riley, Steve Jarmain, Phil Flint and myself, were to fly Gnats. Ian Davidson, Chris King, Nigel Voute and John Bartholomew were chosen for Varsities and Bill achieved his wish of flying helicopters, commencing what was to become a distinguished helicopter career.

The RAF's take on helicopters was interesting. It was a posting requiring high captaincy scores. Flying tough soldiers into bush and jungle required strong captains who would not put up with any bullshit. As for me, I couldn't wait to get my hands on a Gnat at RAF Valley. After that a hill to climb awaited us at weapons school before we were fit to do anything. Meanwhile, I had a mountain in Africa to climb.

There were eight of us on the mini-expedition. The CI Wingco Clarke, Ivor Gibbs, Al Curtis and Brian Penton-Voke plus four graduate pilots, Phil, Riles, Ian and myself so a good mix. We planned to fly out and back on RAF aircraft flying to Kenya. If the return flight was cancelled you pay for civil airfare back. Ivor starred. We had slots out from RAF Brize Norton in Oxfordshire and Ebo tours in Nairobi confirmed two Volkswagon mini-buses which were ideal for dirt tracks.

Mount Kenya has several peaks and I felt the best peak to climb was via the less technical Lewis Glacier to Point Lenana. It was still a challenge at a whisker over 16,350 feet and a Paratroop team recently training in Kenya had failed to get everyone up. The other peaks: 'Bation' at 17,057 and 'Nelion' at 17,021 feet were too technical in my view for novices.

Lenana is more a high trek than a climb so the equipment was simple.

Old fashioned long-handled ice axes were ideal for use as a walking aid and excellent for 'front braking' if you slipped on the ice. Made good splints too but I didn't dwell on that. These, plus a 150-foot half-weight nylon climbing rope to secure any tired guys to me were all we needed.

Arriving at Embakasi Airport, Nairobi, the familiar earthy smell of Africa greeted us, our vehicles awaited and buying rations easy. We drove north up the Rift Valley to Naivasha, then east through the Aberdare Forest where I planned to drive up to the highest place possible and make base camp.

Phil was a rockstar! He tweaked our carburettors with the thinning air giving us several more thousand feet to the top of the tree line. From base camp here we planned to carry everything to the start of the ice where we would dump everything for the climb except for my emergency sack.

It was a cold night and I got a few laughs when I insisted on stoking up the fire to warn off cats. I was more concerned about our food stores than us, but…

Not far from here during my time in Kenya, a lion took a soldier from a tent. They had left the walls up for air and the story was the cat took the second guy in which was a jolly lucky choice because the one near the wall scrambled for his sidearm and shot at the lion as it dragged his chum out. The lion fled without a meal and even accounting for army exaggeration it must have been quite a shock.

We had no such dramas but a leopard did prowl outside the ring of tents in the dark hours. I checked his pug marks in the morning when I made our tea. He was a big one.

The guys were not mountain rescue types and apart from heat exhaustion, the only worry was pulmonary oedema. This is where excess fluid in the lungs can be brought on by exposure to high altitude and is easy to spot by a rattling chest noise. The best treatment was to keep walking down and get oxygen quickly. The Mountain Club of Kenya kept a few bottles for emergencies and I had their positions marked so we were good to go.

The guys found it tough passing the twelve thousand foot mark but cheered up when we glimpsed cheeky rock hyrax which looked like large hampsters sunning themselves. The scenery was spectacular and ahead of us, the peaks of Mount Kenya began to loom higher when the glaciers and steep gulleys took shape it was easy for me to plot our route up.

Wingco Clark showed great enthusiasm and had to be slowed down to avoid altitude sickness and apart from roping two guys on at one exposed ice stretch, we were fine and everyone made it up. I was grateful for the way everyone pulled together as a team with immensely good humour making it so much easier for me.

It was a uniquely private moment for me on the summit gazing southwards where Kilimanjaro's icy dome thrust into the sky. It was lost in a distant haze and I reflected that a few years ago I had sat on that summit with aspirations of becoming a pilot. Now, here I was two hundred miles away on the second-highest mountain in Africa with a bunch of guys who made it come true. I was one lucky chap.

Time is pressing on a high mountain with a long descent so I suggested we start the move down at a steady pace. A twisted or broken ankle meant longer in the cold at high altitudes. Not good.

Before darkness set at base camp, we followed elephant footprints sunk in black oozy mud through an open forest to scrub away days of dirt in a cold stream. I discreetly checked a pile of dung to make sure they weren't warm. Trampled buck-naked by an angry elephant wasn't part of our itinerary.

The next day Wing Commander Clark checked in with the RAF by phone at a country club and it was bad news.

Elwyn Bell and his student had been killed in a mid-air collision with a Naval Sea Prince aircraft over the Vale. No survivors. Elwyn was popular with everyone and we were saddened by his death, but something brightened life for us.

A friendly Commissioner of Police remarked that a delightful old chap lived on the slopes of Mt Kenya called 'Daddy Probin'. He built small aircraft in his back garden and flew them over the bush. Last time out he had lost a canopy over Mt Kenya itself.

Clark remarked he had won the Probin Golf Trophy at Cranwell – could it be the same chap?

It was! Daddy Probin invited us for tea in his enchanting cool bungalow where an old hat-stand caught my eye, it held ancient butterfly nets, golf clubs and battered walking sticks. He lost no time in showing off his latest home-built aeroplane. His staff were all trained in ground handling and he started up and taxied around the garden. It was surreal.

Sitting comfortably on flower print sofas sipping tea from delicate bone china we chatted about some of the old photographs adorning the walls. One showed him drinking beer with a pretty lady on a manicured garden lawn with an old bi-plane in the background. He had run out of fuel on a Cairo to Cape Race and lobbed into someone's garden for a spot of help. I was intrigued about navigation and when I asked was it difficult? He laughed and said he ripped pages out of a Times World Atlas and set off following the Nile. A question of heading south until he ran out of Africa.

Meeting him lightened the mood and we all felt better. It had been a good mini-expedition and after a few days off in Nairobi and a run to Mombasa, we clambered on board a Hercules and Ivor ripped up our cheques.

Bring on Advanced Training. It was time for high-speed Gnats and punchy instructors.

chapter Seventeen

Advanced Flying Training

EXACT POSITION, noisy end of RAF Valley Officer's Mess bar.

After being together for so long we might have been apprehensive about meeting Cranwell graduates joining us at 4 FTS. But we were so pleased to be on fast jets we could have been joined by chimpanzees for all we cared. Happily, Gus Crockatt, Neil Matherson, Derek Poate and Nigel Tingle, were as keen as we were. We had reason to walk tall flying the Gnat. It was a combat aircraft sold to Finland, Yugoslavia and India as a lightweight swept-wing fighter. The Indian Air Force was so impressed with its agility that they nicknamed it 'The Sabre Slayer' in the '65 India Pakistan war.

Squadron Leader Boz Robinson our new squadron boss being an ex-Hunter pilot led from the front arriving with a noisy bunch of instructors to welcome us. He had a certain style about him which I liked. Nursing hangovers the next morning a brief spasm of optimism we would do a bare minimum of ground school died as soon as we heard the syllabus. None, except the Cranwell guys, had bargained for a shed load of advanced aerodynamics. I assumed we would have an hour or so of lectures with maybe a few shock waves drawn on a whiteboard and a few rules to remember: like not doing a sonic run overland near a turkey farm or old folk's home. That sort of thing.

But we had to know much more. And they sold it to us by saying we would be firing supersonic missiles. Not only had we to understand missile diagrams with lift and sonic drag arrows plastered all over them but we had to know what was happening to whatever we were flying if it was supersonic when you fired one.

It was a fair point and I became interested in all things aerodynamic and supersonic. For a brief period, I was on the threshold of becoming technically brilliant and even started sitting in the front row of lectures.

Joining Valley for training were two foreign nationals. They would fly Hunters on 3 Squadron instead of the Gnat. One was a tall cheerful Chinese pilot from the Singapore Defence Force called Burtram Yong who had graduated top of his course. The other was also an immensely cheerful pilot from the Pakistan Air Force. His name was long and I called him Sid. This was not only to shorten it but he and I got into a lot of trouble socially and he is probably an Air Marshal now so it's probably for the best.

Burt and Sid possessed a sense of humour that was even more British than ours and quickly became inseparable companions during happy hour every night.

Technical lessons on Gnat systems swept past like bad hallucinations and they were so fiendishly complicated that I tried moving to the back row for inspiration – but it was full. The aircraft was small, approximately 31 feet long (without a pitot probe) with a span of 24 feet. Powered by a Bristol Siddley Orpheus engine with 4520 lbs thrust, it only weighed 8650 lbs so it was a nippy jet.

It was transonic with a max level speed of 525 knots 0.95 Mach at a low level and went supersonic easily in a dive to about 1.3 Mach. This is 1.3 times the speed of sound, so ideal to demonstrate high speeds to the great unwashed.

Advanced flying training for potential fighter pilots naturally requires flying a high-speed aircraft. This focuses attention on a visual lookout in a quick-moving environment and coping with higher fuel consumption. The level of awareness required was at a higher level than in basic flying. So in other words, you had to be sharper and the aircraft chosen had to be challenging to fly but still be manageable to a dunce coming from basic flying.

The RAF thought the Gnat was a very good aircraft for this. However, the accident rate was a bit high…

Frankly, I was not surprised. A lot of complicated systems were shoe-horned into the fuselage to make it all work. The feel system and powered flying controls I never understood fully. The hydraulic system in the most complicated longitudinal control system ever invented often failed. And it didn't stop there.

The cockpit was small with limited stick movement. And because of a large centre of gravity and trim change when the gear travelled – a fiendishly clever system made the stick remain central when you lowered the landing gear so you didn't run out of space pulling back and spear into the undershoot.

The 'all flying tail' was big-time for me. I even achieved a personal best staying awake for manual reversion in the event of hydraulic failure. The whole tailplane slab would freeze in position with a hydraulic failure but control could still happily be achieved by unlocking small elevators and then moving the slab slowly with a standby electric trimmer.

The only recognisable item was a bicycle chain used in the landing gear circuitry for the ailerons. When clean, travel was restricted to twelve degrees to avoid overstressing but with gear down the full travel of fifteen degrees was available. The rate of roll became mind-blowing when the fuse protecting the circuitry blew. Or got taken out! The ejector seat was lightweight made by Saab and after the seat drill, we fitted out with 'g suits' or turning trousers. These looked like cowboy chaps which inflated hard across the stomach, thighs and calves to stop the blood pooling and helped prevent blackout under g force. My instructor Simon Bostock was an ex-Lightening pilot, and here again, I was lucky with instructors because he had a great reputation for being punchy but fair with a formidable record for getting guys through. A thoroughly likeable person and I never changed my opinion throughout the pressures of the course. Owed him heaps.

The first flight in a Gnat is unforgettable. A step up in every way for us, the lookout had to be better because the silhouette of the Gnat was smaller and moved at a higher speed. We also gobbled fuel at twice the rate we were used to and we travelled a lot further so constant orientation was required so you didn't set out in the wrong direction after a manoeuvre.

The aircraft was so small you felt you were strapping it onto you; the cockpit was snug and when I hit the start the Orpheus engine had an unmistakable growl that didn't belong to a pussycat. It settled down to a satisfying rumble. Loved the noise and vibration. The ground crew removed the Palouse air starter and we were ready to go. Sitting so low with such an excess of power was new, the taxiways felt wider and when we sat on the runway threshold the piano keys on the end looked enormous.

This had to be a tense moment for Simon to take out a new kid on the block but he was cool about it.

Vital actions.

Clearance. Brakes holding at 90%. Tail trim 6 up – Go!

Full power. Engine good.

100 Knots in no time – acceleration check – nose wheel off.

Seconds later 135 knots airborne – brains left behind – gentle prompt from Simon – Gear and flaps up together. Done!

Gentle prompt – 'Wyn, a bit more nose down.'

The view out of the cockpit was beaut' and at 370kts climb speed Anglesey became smaller. Some wag described the view with the long pitot sticking out the front as like riding a rocket-propelled broomstick – and it was charmingly accurate.

Banking steeply onto the climb-out radial I feel the Gnat's agility for the first time. Initial climb rate is 20,000 feet per minute and drops back. but we could still make it in about 3 and a half minutes.

Simon was used to supersonic climbs in a Lightning so to him it was pedestrian stuff. Just a few minutes after take-off, we slowed and Simon pattered me through handling at 300 knots. High angles of bank needed and above 2g my g-pants inflated. Next, acceleration to 500 knots and I could hardly believe I had to throttle back to hold the speed. A real pocket rocket.

Simon demoed twinkle rolls. The roll axis went through my stomach and the rate of roll felt astonishingly fast. I just had to do several before slow-speed flying. The behaviour of the Gnat at the stall was common to all swept-wing aircraft, a flat 'lift curve' and a sharp drag rise at high angles of attack, no sharp transition from un-stalled to stalled like our cuddly JPs. In our hot-rods, we gradually transition to the stall which is perfectly controllable even by ham-fisted pilots like me. Stall buffet and vibration are there but they became heavier and heavier with bags of noise thrown in right up until you fully stalled it.

Then all hell breaks loose, control is lost and autorotation starts.

If you don't get smart with low speed on the dials your heartbeat rockets and you had better look at the rate of descent. It gets very high and to keep you on your toes the full flap vibration can mask the stall buffet very nicely.

If you don't smack on full power, level wings and relax back pressure to a nice climb attitude you will be attending a one-sided interview with the Boss about ejecting over Snowdonia. That's not a good position to be in.

The only worse position is doing this inadvertently on approach and consequently not being around to have an interview with the Boss.

We did every stall exercise possible, having stacks of power is good news and I was reminded to check the fuel constantly. Fuel weight was a sizable portion of all-up weight and using most of it gives a subtle change of handling.

Time to go home. Tacan (tactical aid to navigation equipment) could offset up to 200 miles from a beacon, at Valley, a dive radial ended at a point where ATC picked us up. It's an arrival funnel keeping us clear of other traffic. Diving steeply we tried the airbrakes. These were noisy and fairly unique insofar as the landing gear is partially lowered to act as a speed brake.

Our visual approach was a run-in and break at 320-360 knots at a height of 500 feet and is considered a tidy arrival typically used by combat aircraft to fly at a compromised fighting speed right up until the last moment. Simon demonstrated. The Gnat ate the miles and close to the ground it felt as if we were dancing in the air to the tune of light turbulence. With nothing more than fingertip control, we curved towards the runway where the grey surface had a metallic sheen from a recent shower.

12 Miles – '2 minutes run and break.' ATC 'Roger call 30 seconds.'

3 miles – '30 seconds.'

A climbing left turn from the dead side with the throttle closed and full airbrake to slow us.

Piece of cake!

Speed below 250 Vital Actions. Gear and Flap.

Tight turn onto finals lining up with the runway. Full flap. Power 55%.

200 feet 150 knots, aiming for the 'piano keys'.

Late check: 'Feet off brakes.' I had the impression we were playing darts with the aircraft.

Bang – wheels on. Full power – kick up the backside haul off at 130K almost immediately.

Gear and flaps up. Throttle back 200 knots and 55 degrees bank turn downwind.'Fancy a go Wardy?'' 'You bet' – 'You have control.'

We had a tail chute to slow us down, it's most effective at speeds over 100 knots so you streamed it as soon as you touched down. Once clear of the runway it was a simple matter to turn into wind power up to 65% and the jet blast kept the canopy inflated until you jettisoned it. There were rules and cross-wind limits which I sometimes forgot but I liked the show-off side of it.

That night the bar was full of guys who could scarcely believe the Queen was paying us to do all this. And she gave us weekends off.

Only an outbreak of collective insanity would prevent us from enjoying RAF Valley. The sandy beaches of Anglesey close at hand and the purple-grey mountains of Snowdonia looming across the Menai Straits made it a magical place to be.

What we didn't need was to lose an aircraft practising emergencies. Some drills are inherently dangerous when coupled with another so early flying was a mix of aerobatics and handling to get a good feel of the aircraft before starting serious stuff. Instructors were relaxed but ever-watchful in case we fouled up drills that were easy to do badly – and eject. I daresay there were times when Simon fervently wished he hadn't asked me to attempt the drill in the first place but I detected elation in his voice demonstrating loops, slow- rolls barrel-rolls and twinkle rolls. Speeds were awesomely faster.

My immediate impression pulling up at 350 knots with 4g and tight g-pants was that I was driving it around the loop. No tucking in. Too damned fast. Keeping 4g with gentle pressure back on the stick, vibration as wings start nibbling at the stall buffet, 200 knots at the vertical. Stick coming back – heavier buffet nibble, a plunging 4g, coming down. Super! Straight into another.

Dive down, pull for vertical roll upwards at 450 knots 5g with pants digging hard into the stomach, into the vertical, stick forward with full aileron, a blue sky with raggedy white clouds rotate rapidly past the windscreen. 200 knots stop rolling leave the vertical and pull over the top on the buffet nibble – can't resist an aileron roll on the way down – 'watch the bloody g Wardy!' 6g rapidly on and a badly flown Derry Turn to reverse direction, Simon flew a perfect one to put me in my place. 'Just one more Simon'

'Check Fuel Wardy!' Holy moly! OK, 'Tac' Dive to visual then' 'That's better lad.'

We used adjacent RAF Mona airfield as a relief landing ground which added spice as it wasn't well lit. Time for emergency procedures: I swotted hydraulic and control problems in my bath and Simon re-capped. We covered engine fires and flameouts in the simulator but needed to tackle hydraulic and control problems airborne to cope with forced landings.

The big bogey was the tail. It needs hydraulic and electrical power for normal operation and has a standby feel trim plus a normal trim. If the generator has blown up you can unlock the elevators and operate in 'Follow-Up'.

If you flame out but are the proud possessor of a windmilling engine to give you hydraulics you could be a hero and impress the Boss force landing somewhere in follow-up. It means you don't have to eject and the Boss knows you can read a checklist.

If you lose an engine plus the hydraulics and end up in manual; then you should be walking in with a parachute under your arm because the rules said we must eject (decisions become less tiresome when the last part of the checklist tells you to do this) The instructor turns off the hydraulics from the back seat to simulate a hydraulic failure and this is normally done in the middle of an aerobatics sequence. Straight and level flight would have been far too easy. The drill had to be instinctive and there was a useful mnemonic to use with a hydraulic failure - STUPRECC - which is emblazoned on the soul of anyone who has flown a Gnat.

However, the RAF is by instinct optimistic and gave notes to sort every conceivable problem which we could recite verbatim with high hopes of not fouling up.

Flameout patterns with a wind-milling engine were exacting. At night doubly so. If your engine wasn't wind-milling with enough RPM to keep powered flying controls then the RAF invited you to eject.

After an emergency check-ride, I was ready for solo …

chapter Eighteen

Supersonic. Night Flying & Tragedy

EXACT POSITION upside down overhead Lake Bala

It was June and the weather was stunningly good. Wales lay below in a patchwork of walled pastures, high moor and granite mountains and I had to look up to see it as I was upside down on my second solo in a Gnat. I may have known exactly where I was over the lake but up here I was lost in my little world.

I had a Tacan which told me where I was and I could even calculate the time to get home and get the fuel right. Shame it wasn't working.

But hell, who cared, I carried a dog-eared map clipped to my thigh and Lake Bala was below. Rolling upright I spotted a fast-moving contact and seconds later I picked up another. Two Buccaneers moving like hell using the lake as a target. Tipping into a dive they came off target going wide into battle formation keeping low with a speed of some 500 knots. No doubt they spotted me because they disappeared high speed down a valley and were gone and had I given chase I would have run out of fuel trying to catch them. It was the shape of things to come and I felt a tremendous lift in my spirits. Just a matter of getting through all the courses. Bring it on!

With emergencies with practice forced landings (PFLs) under the belt we developed our handling skills for smoothness in flying. Finesse was required at a high-level limited by our power but at lower levels, it was constant high g for max rate turns.

Sid and Burt were doing the same thing flying Hunters which led to spirited banter in the bar as we out-bullshitted each other. Every evening I had taken to playing the Light Cavalry Overture at full blast when scrubbed up and ready for dinner. On cue, Burt would appear at my door with a huge grin followed closely by Sid and we set off together to join the others in the bar.

We would fly several high-level flights and then do a sonic run to look at supersonic handling and fling in some high-level aerobatics. The lookout problem was now something else. Closing speeds were higher, the manoeuvrability less and the glare from light reflecting off clouds and empty field myopia combined to make our lookout less effective. The myopia part is where the eye, if not looking at something, tends to focus just outside the cockpit a few feet away. To combat these issues we wore our tinted visors down and focused on the horizon. It was cool dude time!

The next day a supersonic run out to sea after high-level work. Our training was thorough. It has to be. The RAF had vast experience of over-confident, hot-headed, ham-fisted students like me.

We started the bang run from 45,000 feet off the Lleyn peninsula. Quick roll inverted and pull through to 30 degrees nose down. No trimming, we accelerate quickly to .98M the needle hesitates momentarily then jumps to 1.02M and my supersonic body dives towards the Irish Sea. Fling in a quick roll and turn.

Ailerons are very effective supersonic – a joy to handle but the pitch stick forces were heavier in supersonic turns. If we didn't have this arrangement the seafloor would be littered with the wreckage of Gnats flown by ham-fisted students like me on a solo boom run. Another mile-stone passed, coffee, de-brief, and a quick reflection on how terrific life is and out for a solo run. Starting the boom dive I was proud of myself for fighting down a strong urge to try a supersonic loop for as long as I did. And I'm not surprised It's forbidden for students. One minute you are supersonic then subsonic then supersonic again! Going through umpteen trim changes made life difficult and rocketed the heartbeat. The problem was coming out of it with a 50-degree plus angle of dive trying to avoid overstressing by not pulling too hard but still hard enough to avoid plunging into the sea. Frankly, I was twitchy but consoled myself I was authorised for aerobatics and a boom run; the only problem comes when you try both at the same time.

We readied ourselves for a tough Instrument flying phase where the pressure was on. One of the Rating Examiners had a reputation for being flint hard and mean. Trip over him and you said bye-bye to fast-jets – just by walking into a room he would drop the temperature and Phil developed a nervous stutter just thinking about taking the test. Huge relief when we all got through and started night flying.

A Gnat's cockpit is low to the ground so taxying at night is entirely different to taxying an aircraft with a higher cockpit. A brightly lit airfield with rain bucketing down on a curved canopy can be confusing because light reflections tend to converge. Despite having different coloured lighting for taxiway edges and perimeter tracks plus a well-lit runway it's still easy to screw up and get lost on the ground. Especially if the speed builds up and you spend more time than you should be squinting at a soggy map clutched in a wet hand. Reversing isn't an option.

Cockpit student folklore was crammed with night incidents – one chap from a hot Middle Eastern country, who was unused to rain never mind taxying a powerful Hunter on a wet night, found himself on the grass. His bright taxi light revealed a stone wall a few feet ahead which definitely should not have been where it was. Well not according to his map. He skidded to a halt mere inches from an infidel wall, shut his engine down, stomped back to the crew room without telling anyone including ATC and fell asleep contemplating who to blame.... And so on!

A first-night climb-out in a Gnat is quite hairy. Highly sensitive controls and quick responses require eyes to flick inside more often to scan the instruments. At a high level, we horsed all over the sky to get the feel of high roll rates in the dark and experience disorientation with the stars. To be honest I found the whole thing sporting which was just as well because the instructors started pulling hydraulic failures to add to the fun.

Every Gnat pilot would agree that a night circuit in the driving rain at poorly lit RAF Mona our satellite airfield was challenging. Flying with a simulated hydraulic failure moves 'challenging' up-scale to a real heart pumper where the instructor in the rear cockpit with his severely restricted view had every right to feel terrified sitting behind us. A relief for them when we moved on to navigation.

We had no gunsight. The RAF wasn't that stupid. Playing with it at low level would have written off half the course.

Briefing points were interesting: "The altimeter is unreliable chaps – just eye-ball it – 250 feet indicated is 400 feet true so get down there".

Fuel consumption is high the weather is an enemy and a visibilities of 3 miles is our limit.

At 250 feet turbulence can be severe. Don't whine about it.

Lookout is essential to avoid terrain, birds and other aircraft.

500 knots is the fastest we would fly, this is a high rate of travel over the ground so intensive lookout, It's difficult to map read in turbulence and avoid hitting the ground.

Bird strikes cause severe damage above 400 knots minimum height is 500 feet to start with.

Turning circles are large. Take a racing car line into bends in the valleys.

Height is a visual assessment 250 feet is approximately three times the height of a normal overhead electric cable. An emergency pull-up is a rapid 30 degrees climb at a minimum of 360 knots at full power.

Ground to 5500 feet covers two miles forward and you reach it in seconds so think ahead when the weather goes to worms. And don't even think about turning hard inside a valley to go back the other way.

Engine failure - pull up; eject and don't argue.

The Hunter boys did the same but practised low-level turns inside a valley.

We would do this at RAF Chivenor where it was regarded as challenging flying. Just how dangerous was proved when Burt took off for some dual instruction on low-level escape – but they never made it out of the valley.

They hit the ground at Devils Bridge and both Burt and his instructor were killed. It was a rotten shock as he was a really good pilot much respected by us all. He had been close to us as a comrade and we felt sorry for his parents who we knew were extremely proud of him. I never played the Light Cavalry Overture again before dinner. Didn't mean the same anymore without Burt. Sid confided he felt the same.

However, something cheered me up. The Red Arrows, the RAF Aerobatic Team flying red Gnats arrived at Valley for annual simulator training. Flying at an impossibly low height trailing smoke they beat the place up. It was awesome, their panache and casual professionalism in an impromptu display just blew me away. The following day we started formation flying with a measurable spring of over-confidence in our steps which added to our instructor's apprehension. This part of the course was high-risk enough without us seeing the Reds in action!

Formation briefing for the first sortie took longer than the actual flying part and was as interesting as it was deadly serious. A simple error could unravel into a nightmare scenario in seconds which is why I suppose one or two people looked a bit bunched up.

With my over-enthusiastic approach to flying I was accustomed to getting out of trouble I shouldn't have got into in the first place so I was reasonably relaxed. The format we followed was familiar with all fast-jet pilots launching off as a pair.

Domestics – frequencies, recovery and fuel calls.

Fuel calls are vital – a 'Bingo'call meant a predetermined fuel left and if the Leader was organised he was heading home it came. 'Joker' fuel was the lowest state and you needed to be home soon after calling it.

The Gnat with its sensitive controls and pokey acceleration might be challenging, and the instructors wary, but there was no mistaking their elation in teaching us formation. Flying as a pair is less embarrassing than a VIC of three when you foul up. This is easily done as early as your first take-off by over-anticipating the Lead's hand signals. Setting 90% power he chops his hand down and releases brakes but you are spring-loaded and just a bit cocky. A split second before he chops down you have already released your brakes and are forging ahead almost colliding with his wing. You promptly drag the power off, he accelerates away and you slam the power back on but it's too late and you trail miserably behind. I caught up after an appalling takeoff trying to ignore the guffawing from the backseat. It's now a world of delicate, almost fingertip control concentrating like hell on sticking close but trying to be aware of everything in the sky.

Easing out to practise join-ups we hang like swallows before dipping back into formation. We swapped positions. We 'broke off' to simulate bad weather and re-joined quickly, keeping lower than the Lead so we didn't hit him. It's sweaty stuff and a lot of nervous energy goes into it.

But! I grin behind my mask the whole time.

Tail chasing is Intoxicating. The sense of freedom sliding across the sky, skirting clouds and banking steeply in the blue gaps is exquisite. I wasn't an ace or an angel in after-burner – I was just in another place. I discover that when solo you could break all the rules if you were at the back. Just had to keep out of sight in the lead's mirror and try not to frighten myself by getting too close.

chapter Nineteen

A Dining Out. A Strip-Tease Dancer.

EXACT POSITION 3 nautical miles south of where I should have been.

Pin-point navigation at low-level wasn't as easy as I thought. I had confidently chosen a small croft-like building as a target to impress everyone and having cocked up my timing and missed it I had to endure a rising level of mirth from the backseat.

After refuelling at RAF Lossiemouth, taking good advice on the return trip I overflew huge stately homes and if you discount the low flying complaints, it went well. Two days afterwards, I managed to whizz through the final check and in short order, so did the others. We all passed.

Our instructors were as eager to celebrate as we were on this belter of a night. One where champagne bottles pop, swift-eyed stewards kept our glasses full while we all back-slapped and made vigorous handshakes. As a multitude of friends and colleagues pumped my hand the realisation that my journey from a saddle to a cockpit was now fully complete raised a lump in my throat. In a private moment, I felt a deep respect for the RAF for making it all possible. It was that familiar feeling of being part of a family again. A feeling that had carried me through some jagged moments.

My wings were permanent. Unless I did something stupid…

It wasn't all my fault. No, the fault lay with the ill-advised free beer put on by the graduating macho-men preceding us. When the graduating course before *theirs* provided two free barrels of beer to thank everyone. It had to be outdone. So they provided *three* free barrels of beer to make it more memorable.

It was a real dilemma to even think about *four* barrels. No one in their right mind could drink that much beer. An all-hands meeting in the bar annexe was called.

I sat in an ill-lit corner listening to awful ideas ranging from providing four barrels of beer, playing Tom and Jerry cartoons all night to hiring a magician.

All we needed was a choir of parrots singing all things bright and beautiful.

It was an oblique reference by one of our Cranwell contingent about respecting RAF ethos that drew me from the shadows. You see, I knew a thing or two about 'Ethos'. The ancient Greek word for character, a moral character that bound a community together (or squadron in my case, whenever given a dirty job I was always told it was all for the squadron's bloody ethos)

Many guys from my best-forgotten past certainly possessed character – I'm not so sure about morals. But that's beside the point. I know they would create social mayhem on such a night. Sid was in total agreement with me hiring a stripper.

And she was going to perform on the head table. The Linton guys who followed me through the African bush and mountains liked the idea of following me into trouble as did Gus and Neil. Some of the Cranwell guys looked apprehensive until I pointed out that after speeches the Station Commander always departed to the ante-room with distinguished guests for a brandy before the unwashed arrived. This left us, the odd Boss or two and the Padre to port and cigars – and that's when naughtiness would strike.

I avoided: high-kicking long-limbed beauty and the naked word in case I gave the wrong impression to a Cranwell persona. 'Naughtiness' was the right choice.

What's more, in the unlikely event of someone complaining I would take full responsibility.

We hired a charming artist and booked her into a nearby hotel. We even sent her a map of the dining room with the Padre's seat marked with an 'X'. With the important planning done for our immediate future, it was assessment time for our long-term role. The deciding factor would always be an empty cockpit. It could be in the Buccaneer, Lightning, Phantom or the new Jaguar force. The roles, broadly speaking, were ground attack or air defence and this is where the experience of the instructor's reports lent weight to someone's choice. If you had done well, you got your first choice.

The career-minded preferred to stick with one role for as long as possible. Gaining experience they shone out from the herd and moved to greater heights so the role was immaterial compared to its longevity. This didn't apply to me, I sensed bucket-loads of fun over the horizon and any vague ideas of fitting in a career of sorts would just have to wait.

With a natural inclination towards legitimately blowing things up and having a good scrap in the air, something bloody fast low down and heavily armed was a good choice.

One aircraft fitted the bill beautifully – the Buccaneer.

The Navy had RAF guys flying them from aircraft carriers, and their crews had a reputation as highly-disciplined apostles of low flying (I also heard some were right rip-shits at it) A perfect choice.

Young Riles was a sharp pilot and he wanted the Lightning as did Neil, Steve J and Gus. Phil ended up on Jaguars so he probably mixed his choice up with his cars and I wanted Buccs with a Lightening tour afterwards so I didn't miss anything.

The theatre of dining-out nights draws me into a relaxed mood. Dressed in mess-kit strolling under chandeliers raising a glass of sherry or beer to squadron friends is a good feeling. One of appropriateness if I'm honest. But this particular night despite several glasses of sherry I felt a niggling apprehension. I put it down to the disturbingly attractive girl who would slip into the mess-secretaries office at the top of the stairs. They led down to the dining hall where I would meet her at a precise time. She had giggled when I insisted on synchronising watches.

No. It was the guys who kept sidling up to Sid and me, slipping notes into our pockets murmuring 'Good show guys, a small contribution,' who made me apprehensive. I hadn't told anybody! Word was out.

Our distinguished guests stood at the head table, the station commander called for grace and the Padre obliged. He kept it mercifully short and judging by the serene ecclesiastical expression he wasn't aware of shades of Sodom and Gomorrah to come.

'Too late now Wardy', I sighed to myself, settling down to a large glass of white.

It was an auspicious time for our course. My posting was confirmed to Buccaneers to wild cheering from friends and instructors. And it was evident we had all done well because everyone got their first choice to loud cheering and table-thumping. It pleased the station commander too. After speeches, he led our distinguished guests to the Ante-room with a beaming smile on his face.

I didn't see him go because I was hiding around the corner at the top of the marbled staircase with a lively stripper on my arm.

As venues go, this was terrific. Soft light from glittering candelabra falling on squadron silver scattered over acres of dark oak tables – and Sid winding up the gramophone. To the strains of 'House of the Rising Sun', our lady sprang elegantly onto a table and amidst wild cheering seductively danced her way towards the top table. I stood alongside and every time a silk veil came off, she tossed it to me and I would catch it to wild table-thumping and drape it casually over my shoulder. By the time the seventh one came my way the cheering swelled to a wild crescendo. Playing gallant for her modesty, I tossed back the last veil and we legged it up the stairs.

My timing was unexpectedly brilliant. The dining room door crashed open revealing a smouldering Group Captain. No smile on his face now. Just a shaking fury in the way a volcano does when it blows. The cheering died.

He roared out: 'I'm going to court-martial the officer responsible for this disgusting behaviour in the mess'… 'You are supposed to be officers and gentlemen.' I couldn't remember what else he said but it was stormy stuff.

Three squadron bosses, a padre, and several squadrons of students and instructors knew exactly who was responsible. I was standing at the top of the stairs edging slowly into the mess secretaries office covered in silk veils with my hand over her mouth to stop any giggles.

After seeing the lady off in a taxi I spent most of the evening wondering ruefully how long my posting to Buccs would be delayed by a court-martial. Most instructors said it would improve my chances with the Royal Navy which cheered me up, things couldn't get worse.

The next morning is best forgotten. I spent a lot of time touring office waiting rooms, starting with Boz, my squadron boss. I said I took full responsibility, with no one else to blame. It was meant to be a bit of a surprise. He kindly informed me it certainly was but warned me the Chief Instructor wanted a word. And he wasn't a happy bunny.

I said the same to the Chief Instructor but he wasn't so kind. He was seething and told me the President of the Mess Committee who was Wing Commander Admin wanted a word. He turned out to be even less kind and kept me waiting longer. But he liked the fact that I owned up to everything because it made a summary of evidence for court-martial so much easier. He said the station commander wanted a word and was still furious.

Sitting in the waiting room I didn't think a conversation in curt tones boded well but when the door opened, I was told by some office dweller to return to my boss's office. Something was up.

Boz informed me with a tight smile, that Sid had sterlingly made the same office rounds admitting full responsibility. That I had been dragged into it. In the confusion of who was to blame, one thing was clear. Court-martialling a Commonwealth National was out of the question. Instead, we were both invited to donate quite a large sum to the station commander's fund. And to get the hell out of there before he changed his mind.

chapter Twenty

Flying Hunters

AFTER GRADUATING without a court-martial I descended on RAF Chivenor in Devon to begin a long day-fighter ground attack course (DFGA). Jerry Barnett, a forthright likeable guy, ex-Cranwell and keen as mustard was the only other pilot on the course.

RAF Chivenor was originally an old Coastal Command airfield on the northern side of the River Taw estuary in the Bristol Channel. Home to the RAF Tactical Weapons Unit it buzzed with activity. Everything was different about Chivenor. Agreeably different. Our rooms were in creosoted wooden huts that dipped in the middle like ailing Noah's Ark. Rabbits bolted in the long grass between them as a friendly batman of some years showed me the way, enquiring politely in a slow rustic Devon accent how I liked my morning tea. He thrust a door open revealing a friendly pot-bellied coke stove burning at one end and a single vintage iron bed stood at the other. Creaks from the hut and the smell of coke from the stove mixed with snatches of burnt kerosene drifting from a pair of Hunters taxying past to the runway. Our conversation paused for a moment as windows rattled to the rolling thunder of their Avon engines at take-off power. Departing swiftly they left an empty silence and the smells of my new home. Brilliant.

I noticed a change in the attitude of our instructors as we shifted higher up the skill chain.

Acceptance in the aviation fraternity is a gradual process and I wondered if this would continue when we loaded up with live ammunition.

I speculated that our instructors here would have to be tough. Mean even. Gun-slinging mean. After all, the RAF letting us loose in a fully armed Hunter was the equivalent of packing an excitable teenager off to school with a loaded shotgun.

But incredibly, when we met them at happy hour none of them breathed fire, had horns sticking out of their heads or carried tridents. They were open and friendly although I noticed they fixed you with a direct look and their pleasant punchy manner suggested someone at the top of their game.

The technical side was uncomplicated and we quickly hoisted in all the cockpit drills Camouflaged aircraft armed with live weapons lining the bays gave a perceptible feel of a fighter field. Pilots often commented on the atmosphere at Chivenor, the helpfulness of the place and the furtherance of fighter ethics.

During coffee breaks, the conversation leaned towards world war fighter aces. The top ten accounted, conservatively for over 300 certain kills between them and that's not counting damaged and shared *kills s*o you could double that figure and end up with a sizable number of German squadrons shattered by just a few guys. They could shoot!

At this stage, we budding aces couldn't even fly a Hunter never mind shoot at anything. But that changed on a spring day in March when I strapped into a two-seater Hunter T7 with instructor Peter Millard.

Pete was likeable calm and experienced and ex- Middle-east and put me at ease. Sitting in a cockpit with triggers and firing buttons for rockets, bombs and guns were novel but no time to dwell on it because Pete had strapped in and was politely drumming his fingers on the stick waiting for me to catch up -– I hurtled through the checks and when I finished he gave a wink and I knew I had learned my first lesson. A Hunter T7 is a bit cramped sitting side by side but straightforward to fly and in no time at all it was time for single-seat F6s.

By now I had picked up the practice of kicking the nose wheel of any new type I flew for the first time and growling: 'OK remember I'm the b****y boss' before strapping in. It felt even better now with four smoke-blackened cannon ports just above my head.

Lining up on the runway I received takeoff clearance and gave a loud yiiaaah! went to full power released the brakes and straightaway knew this was a thoroughbred. Acceleration forces me to clean up the gear and flaps smartly before barrelling out over Saunton Sands. Seconds later gunmetal grey waves of the Bristol Channel whipped beneath as I revelled in the obligatory high spirits that come with being paid by the Queen to fly a new type of aircraft. Although to be honest I had to throttle back a bit just after taking off to work out what the hell was going on. The F6 was poky. Controls were light and responsive and the acceleration kept on coming.

I headed south towards Clovelly following the coast to have a good look around. The crisp handling of the F6 was immediately apparent as I threw it all over the sky. Powering into the vertical I let the speed fall off as we had on the T7 to see what it would do when I fell out. Recovering from auto-rotation was so easy I changed my mind and threw caution to the wind and spun it. I hadn't been briefed to do that but I couldn't help myself and besides, I didn't think the duty instructor would mind. The recovery was quick and took hardly any height which was a good thing because it would have been a bit difficult to explain my presence in a dinghy in the Bristol Channel (I think I was a little naïve in those days)

It was a super aeroplane, three solo trips in three days throwing it around feeling the power and performance was a joy; conversion and instrument rating took ten trips. And then the fun started.

We could fly in close formation but knew little about battle formation; this is the basic tool used in positioning more than one fighter for defensive cover or to act aggressively. There were two basic formation positions: fighting battle and wide battle. In 'fighting', a wingman would maintain a loose position astern inside a manoeuvring cone about twenty-five degrees on either side of the leader's flight path between 50 and 250 yards. This allowed a good lookout and avoided colliding with the leader who might be manoeuvring hard to get his guns or missile to bear on some unfortunate.

In wide battle, the number two would fly parallel about 2000 yards abeam looking across to cover each other's tail; If attacked, a hard turn sandwich the baddy and expose him to the guns of the other. The abeam distance opened to 5000 yards or several miles covered a missile environment.

If a formation was larger than two aircraft, number three was deputy lead and flew abreast. In a four-ship, numbers two and four flew fighting-battle on their respective leads.

Screwing up a turn in a wide battle formation and not crossing properly earns much wrath, because, it's essential to know who misses who. At high level, it's possible to be close to the stall if you foul up so delicate handling is required; turns are much wider too and the trajectory of the other guy is different. The eye-watering white orb of the sun always seemed to be where I was looking and I held up my thumb against it. Slowly our head movements become mechanical, a metronome regularity. A habit. And my peripheral vision is more acute. Concentration is enormous and I was thankful to plonk the wheels on in a burst of blue smoke for a cuppa. We plan to jump into F6s and roar off to do the same again as a three-ship.

Instructors convey a powerful message at this point. No respectable fighter pilot ambles out for a casual walk around, eases into the cockpit carefully folding his map after combing his hair. That's strictly for 'Truckies'. Jerry and I did a normal external strapping in without wasting time but when I plugged in I heard Lead requesting taxi clearance. Unbelievably, I saw him check his brakes and taxi out before we had even started engines! After that, to be honest, all I ever really checked was the aircraft number so I got the right one, strapping in was strictly on the hoof taxying out.

The dangers of colliding during a robust low-level battle session were pretty high until we got the hang of it I must say. But flying in a four-ship below clifftop level, the lethal shark-like shapes speeding over cresting waves makes you feel you are part of the cutting edge of the RAF.

As was the way at Chivenor.Just when we thought we had something hacked we moved on. Time to use the gunsight!

Cine Weave was a great sport for students but the rules had to be heavily underscored: target fixation and sweaty near-collisions were a double act that cropped up in every course with guys trying to be cool aces.

The gunsight dominated the windscreen and because a junior fighter pilot was bound to abuse it, it had to be simple to operate.

A twist grip on the throttle controlled a ranging ring to fit over the wingtips of the target. One eye saw a fixed cross, the other eye saw a small round aiming pipper, both focused on infinity so we could see both the target and the sight.

The fixed cross aligned with the fore and aft axis and the pipper predicted where the cannon shells would go once they left the muzzles. Initially, they hurtled towards the fixed cross but g force, gravity drop and a host of factors known only to attack Instructors affected the trajectory and they hit through the pipper.

All you had to do was track the target as it manoeuvred, adjust the range as you got closer and closer and try not to collide! In theory, it's simple, there was even radar ranging to do it for you but not allowed to use.

Cine weave was a canned scenario that exploded a few myths...

The aiming point was the pilot's head, which the aiming pipper had to be centred on for a crucial 3 seconds minimum, otherwise, you could not claim a hit. The RAF didn't pussyfoot around training us to be gallant aiming for an engine or something big and then salute the poor chap in his parachute. No Sir, you tried to kill the pilot and blow everything to bits; to help this along the Aden cannons each fired 25 rounds per second and you had four of them.

The attacker called inside a gun range of 1300 yards or so prompting the target to fly a weave manoeuvre. At long range, the pipper seemed attached to a loose elastic band. It wavered all over the place but closer in it was more rigid and aiming easier; bringing with it the problem of hanging on too long to get more film to prove you were an ace.

Unhampered by any skill we 'titled' cine magazines with our names to prevent cheating stuffed them into map pockets and blasted off. Amazingly we improved but had to work at it. During turnarounds, over a coffee, Jerry and I spared a thought for the poor new guys in the Battle of Britain who never had this intensity of air-to-air training many joined a squadron in the morning and didn't even make it to lunchtime.

On low-level navigation we were chased by instructors to mark our accuracy; some targets were one-plank bridges too small for a sheep to cross but we were told to stop whining because it was like bloody London Bridge compared to what was coming on fighter recce'. No time to argue – it was time for gunnery.

chapter Twenty-One

Cutting Edge

30mm Aden cannon

THE RAF can be parsimonious at times but the amount of ammunition used to learn to shoot straight–bordered on extravagance. Air-to-ground gunnery starts with a dual live firing sortie in South Wales on Pembrey Range and following a cine run, live firing. The same for sneb rockets before combined live guns and rocket sorties. Sneb was a versatile French rocket with a high explosive or armour-piercing anti-tank head, they could be fired in a single-shot or rippling salvo of 18 rockets a pod. However, there were limits to the RAF's generosity: we would fire single shots.

Level skip bombing and dive-bombing to follow then air to air firing on the flag. Providing we were still around, the air combat phase started with one versus one, then two versus two which only left fighter reconnaissance.

Range discipline was vital. It was hammered into us that the range officer was God and his gospels included making accurate slot times. Various sins were listed including firing too close in and crossing a foul line to improve scores. If fouling was accidental because it's your first time you got a rollicking and a warning. But if you were cheating and did it several times you were sent home in disgrace because fouling exposed you to Murphy's Second Law: 'what goes down can come up' and subsequent ricochet damage.

A flagrant breach such as beating up the range tower with live weapons was regarded as an unseemly act with dire consequences including an immediate change of profession. Range weather dictated our programme and briefings included another acronym: DWIFOG - don't walk in front of guns. Electrically fired, they had been known to go off once battery power was applied.

Pembrey Range gave us clearance and Paul Dixon talked through a dive onto the target. My pulse went up when he called 'In Live.' Tipping in, the pipper steadied on target I felt a shudder through the airframe and a noise like a pneumatic road drill hammered briefly beneath my feet. A whiff of cordite. Brilliant!

Tipping in for my first strafing run full of expectation I flipped off the safety, and the trigger dropped into the crook of my index finger. Pipper on target I squeezed off a two-second burst to a beautiful metallic hammering beneath my feet. Pulling straight up and grinning like a Cheshire cat I palmed the trigger up flicked down the safety cover and turned abeam.

There were four guns on an F6. They could be fired in pairs selecting 'outers or Inners' or 'All'. The Pilot Attack Instructors (PAIs) warned against firing all the guns together; the recoil was about half the thrust of the engine and cannon vibrations could trip circuit breakers. Only one gun of a pair was fused and we were forbidden to switch to 'All.'

Confirmation of aim comes with the result so it's imperative to know the score. Some targets were 'acoustic scores' where you were informed of hits each pass. Other targets were just screens and after counting holes, scores were passed on to our squadron at Chivenor. Immediate results other than 'on target' were not known until signed back in with the duty instructor who posted them on his wall.

Tension mounts on the way home knowing inquisitive eyes would view your results before you. Striding in from the flight line any generous offer to comment on your score before you've seen them has to be treated with studied indifference. As if you didn't care; but underneath you burned with fervent expectancy.

Next up; – rockets. Dual trip with Knut Fossum a Norwegian pilot on an exchange tour, Knut was a cheerful outgoing character who habitually wore his Dad's old leather flying jacket and started every conversation with: 'you British are sooooh punchy!'

Knut demoed a shot at a circular target of oil drums. Dive angle was similar to guns and the rocket gave a satisfying whoosh with a nice flaming trail leaving the pod but as in cannon fire, you didn't wait to see the fall of shot because of the high closing speed. Great stuff. Over evening beer discussions centred around scores, cockups and comparisons and sometimes I found it hard to get to sleep with the stimulation of the days flying. Despite wearing g pants the constant diving at targets was tiring but a weariness without stress; slumber always came.

FAC (Forward Air Controller) were the guys who told us where to aim and these were often RAF pilots working with the army. It's one of the hardest tasks in the ground attack role. We had to know the FLOT - Forward Line of Own Troops so we didn't fire on our chaps and where the target was. But these simple tasks were not that easy at 420 knots and 250 feet trying not to hit the ground pencilling stuff in; targets were also camouflaged and not easy to see. Great sport but hard work. Luckily the weather proved fair during spring of 72. I hadn't dropped a sortie, only had one gun stoppage and no Sneb misfires. Life couldn't possibly improve. But it did. Time for air combat training.

We demonstrated we could use gunsights. As one instructor put it: 'enjoyed hitting something we more or less aimed at'. The final weeks would see us put it all together in air combat training. But astonishingly, some voiced opinions about doing any of it.

The thrill of flying a highly manoeuvrable aircraft inevitably leads to treating flying like a game. But underneath, everyone was fully committed. In operations in Radfan and Oman, pilots commented on the brutal gap between training and live operation but they still got stuck in.

A crucial part of our training was developing aggression with the right frame of mind. Instructors were proficient at spotting flaws in someone's makeup. And in my opinion fair. Guys who couldn't shoot straight were given more chances but if no improvement came it was a regrettable parting. The same could be said of frequent near-collisions and lack of awareness at high speeds – too many moving parts as it were.

But I remember a character at Valley who habitually wandered around with his hands in the pockets of his unbuttoned officer's raincoat. He seldom wore a hat or respected an airman's salute. He told me his university squadron time allowed him to fast-track, that's why *he* was a flight lieutenant. Such remarks could have raised a boundary between us but they didn't because the ex-university squadron chaps I met in Coastal were great guys and I enjoyed flying with them. I simply felt he had a lot to learn. This is exactly what the instructors at Chivenor thought when he pitched up. Particularly the one he told it was just a tick in the box for him on his way up and it wasn't the wisest choice of words either telling a weapons instructor that he didn't have any interest in all this weapons stuff. Wasn't it-only for show anyway? I often wondered how he felt returning so quickly to civilian life.

A change in my future came In the noisy depths of a happy hour where I was chatting to someone down in his cup*s* because his Phantom course had been cancelled again. I was about to tell him he might be better off because all Phantom pilots drank milk etc when shouts drowned me out.

Speculation that a tour on Hunters would provide a pool of fighter pilots to ease the dilemma of training delays had circulated for months. Rundown 'East of Suez' fed this beast of a problem; more pilots available led to top-slicing courses with good guys wasted waiting for new Harriers and Jaguars.

The guy at the end of the bar had overheard that two Hunter squadrons were reforming.

Monday morning saw me briefing room for air combat training – with 'you lucky bastards' written against our names! Jerry and I were off to the new Hunter squadron at West Raynham. We had to finish as soon as we could. A dream posting.

For embryo fighter pilots the rules of combat training have to be uncomplicated so we could remember them. They boiled down to not hitting each other or firing live rounds at each other. Guns had to be disconnected and taped.

No 'head-on, sight on' passes. In other words, don't play 'chicken'.

Miss by 100 yards head-on, 100 feet vertically.

Stick to a floor altitude, i.e. combat off at say 5000 feet (too much of a temptation pointing vertically down behind someone you worked hard to get behind and ignore the ground hurtling up)

By anyone's standards, this was gripping stuff and before long the board was covered in the baddie's in red and hero's in blue. Main manoeuvres: Low-speed yo-yo, high-speed yo-yo, scissors and high g barrel roll. I knew that a yo-yo went up and down which was a good start. Immelmann and Chandelle's turns didn't figure. The brief was short; the sky was our classroom. Flight Lieutenant Drake was the brave soul who sat beside me at 20,000 feet in loose formation with another instructor. On cue, he pulled alongside and we split 45 degrees away from each other and after a minute or so turned to face.

I was shown the delicacy of fine handling and how lateral and vertical yo-yo worked and Drake kindly threw in how to feint passing the other guy, gulling him into a wrong turn.

I was clearly on the verge of something magnificent here – the sport of Kings. Marvellous.

My immediate impression was that you might sight a tiny black dot at 25 miles or so and think you had plenty of time but at a closing speed in the order of nine hundred plus knots, this speck rapidly took on a menacing look and by the time you were abeam It was best to have some sort of a plan!

The first to sight the other might hold an advantage but it could be thrown away.

Years of experience in air combat training were decanted into the next hour just for me. To get behind our long-suffering target, we had to lose excessive overtake speed. A high-speed yo-yo involves a high arc upwards and then downwards to roll out behind your target and vividly brings home the third-dimensional aspect of air combat. We did the same again but using a high-speed barrel roll to bleed off forward speed. Positioning on the opposite side of a 'circle of joy' we smoothly moved up to gain height and then convert it to speed in a gentle dive down to edge closer and closer...

I tried the 'scissors' where we scissored back and forth trying not to expose our six and force the other chap ahead. Fuel burned off rapidly with no time for repeats and when we landed to refuel we drank mugs of sweet coffee eager to get stuck in again.

What struck me about the physical aspect of air combat training was the amount of time spent craning your neck backwards to look up. Moreover, I felt surprisingly knackered with mental concentration. The instructors were rational, dogfighting per se was fundamental training, a bedrock of skill but missiles and passing bursts of gunfire would be the norm not chasing around a circle. Getting the geometry of interception right would win the game where the future would be head-on from miles away or heat-seeking missiles.

On level skip bombing, we weren't trying to be Dambusters; we just explored the aiming solution for cluster-bomb attacks on tank parks. Dive bombing was a great sport if the angle was steep but brought home the fickle nature of the wind between me and the ground. Although it was fun learning, I was aware of feeling vulnerable to barrage fire during level-skip bombing, and in the dive, I felt even more exposed. Warfare was changing rapidly; electronic jamming was going to be centre stage in a conflict and if the great unwashed got hold of up-to-date missiles who the hell knew what would happen. No point in worrying, because we were straight on to FR. Fighter Reconnaissance (FR)is a role that comes with challenges and the RAF had a successful aircraft in the Hunter FR10. At first glance, apart from the cameras, it looked identical to the FGA9s but inside items had been removed and balanced with an armour-plated cockpit floor and instrument consuls. The camera controls were simple and the view forward better because the gunsight was offset.

Flying the FR10 was easy and at the fighter reconnaissance school, we had some of the best instructors in the game. Dereck Bridge taught FR with Tim Thorne a tall, well-spoken South

African who always seemed to have a smile on his face, and a tennis racquet in his hand. A legendary figure in FR. For a sneaky view of something someone felt suspicious about: first, we had to get there. So accurate navigation. The army, our usual customers were sticklers for timing when hurling heavy artillery shells around the place, so time on target had to be exact. Furthermore, the target was only in sight for about seven seconds, it had to be photographed and described accurately over the radio on our way home. We boned up on all Soviet-made armoured, dams, bridges of every kind and comms sites. And it all had to be described properly with no flippancy.

Thankfully, they didn't believe in miracles. Derick passed endless photographs under our noses for seven seconds, flipped them out of sight and howled with laughter at our description. But he had techniques to help. Using them I thought we might be useful – until a Photographic Interpreter (PI) looked at my film. They miss little and working together they taught us to look *for* something, not *at* something. But no time to dwell on it. Something big lined up – my first tactical two versus two and after that, I was off on a date with a petite young lady in Yorkshire. As in everything we did at Chivenor, you learn quickly, sometimes the hard way – like my first two v two!

First, I learned that manoeuvring hard and trying not to collide with my lead was nerve-wracking.

Secondly. I learned that it pays to be careful with your engine handling when excited. Going from idle to max too quickly is not good. I couldn't see why I was falling behind at first: the lack of engine noise wasn't noticeable because of my heavy breathing but when warning lights came on I knew I was in for a spot of trouble.

My strangled 'Gold four flameout' stopped the chatter, I hit the relight button and turned for Chivenor. No immediate problem high over the Bristol Channel, the weather was great and I could see where the airfield ought to be. Barry Stott, number three was on me in a flash. I confirmed my problem in a calmer voice trying to get an offhand tone into it as if I was used to this kind of thing and pressed on with the drills – Nothing.

With height in hand and ever the optimist, I decided if a cold relight didn't do the job I could still make the hot date if I managed a decent forced landing profile (what goes through your mind when everything goes pear-shaped is amazing) And if I got my ill-treated engine back I would do a flame-out profile just in case. On the face of it, nothing much could go wrong. But I was careful not to even think that.

Barry Stott was brilliant, edging ahead at the correct glide speed when I got a bit sloppy busying myself with various cocks and levers. And it all worked out. Well almost.

My engine belched, rumbled and warning lights went out.

Great! But we were in the flame-out pattern. Didn't need it now. Barry was a star and as soon as I got my turn towards the field sorted he parked himself on me and called I was looking good. This was just as well because I discovered the engine wasn't working anymore. Warning lights came on and stayed on but I managed to get things right so didn't need my engine, Barry waggled his wings and pulled away smartly to get back in the scrap.

It wasn't my best day at work and I didn't make my date because of all the paperwork. But it turned out my next flight was a delivery to 45 Squadron. I enjoyed a demanding course taught by remarkably talented characters with that abundance of good humour which always clings to accomplished guys. To have enjoyed it so much was all down to their fighter ethics and pride in being the best. It was time now to meet my new boss. A DFGA man down to his boots.

chapter Twenty-Two

Hunter Squadron

SQUADRON LEADER Wally Willman re-formed 45 Squadron at West Raynham. Wally was a lively character with a vocal sense of humour and a wicked organiser, a fiend on the squash courts early evenings saw his stocky figure, squash racquet in one hand and a beer in the other finishing off the last of the day's problems. Squadron Callum Kerr was posted in as Flight Commander and Flight Lieutenants Jerry Barnett, Dave Farquarson, Mike Giles and myself were junior pilots. Mike Fernee, Mike Crook, Roger Hyde and Cyd Sowler were the veterans posted in to sort us out.

45 Squadron hadn't been given a formal war role yet but it was expected we would form a reserve squadron. Two squadrons were planned to fly operationally in the DFGA role and act as a shock absorber in the training system by providing a pool of experienced pilots for Harriers Jaguars and Buccaneers.

Wally gleefully told me over a beer that he was determined to raise the level of expertise to that of the disbanded Middle East squadrons. He started straight away. High-level battle formation tactics were at our ceiling which put teeth on edge being close to the stall buffet and bounced regularly to make it difficult.

Standards were exacting and it wasn't training for training's sake, we were being moulded into a useful force.

Reluctantly, we left Raynham in September '72 for RAF Wittering. It was already 'home to 1 Squadron squadron Harriers who lost no time in landing one next to their crew room to crow they were a vertical bunch.

With two fighter types on one station, it wasn't going to be dull. The OCU for the Harrier developed a peculiar language of their own – everything was 'bona'. They were a friendly bunch and in the bar I soon got it all mixed up, for theirs was a language of short takeoff and landings, vertical landings or even rolling vertical take-offs or something like that and when my eyes glazed I just nodded my head politely. But in my opinion, the bona-jet was difficult to fly and easy to crash.

Fortunately, we used the same weapons so it turned out to be 'bona' for us as the place was full of ammo.

In my case, and Dave's, having lots of ammunition was our undoing but it wasn't entirely our fault (gunsight film footage would show it was a competitive thing)

The Wash ranges of Saltfleet and Holbeach were a mere hop away and Wally let us loose on Saltfleet where we overflew the targets and I peeled off to go first. Our rules were to load all four guns with as much as the plumbers could get into them, select inners or outers and fire out in two bursts. Proper strafing – none of your plinking rubbish. This was 'big boys' stuff and knowing how punchy Wally was and how keen he was on accurate gunnery, I thought I would give it my best shot so to speak.

I opened up sighting a tad high squeezing off a wickedly long burst. Too long. Way too long and I think it was seeing the fall of shot for the first time that did it. The targets were large stretched canvas squares and I glimpsed a cloud of boiling sand smack on the target just before it vanished. Wary of fixation I pulled smartly off-target and a second later felt shells hitting me with a series of loud thuds. I guessed I had some ricochet damage which I ignored for a few seconds because the range officer said 'target destroyed take number three next', adding 'getting too close.' I thought ''target destroyed'' had a nice ring to it. Deciding to sort out the damage I flicked the speed brake out and was puzzling out why I wasn't slowing down just as Dave called 'in hot.' I could tell he was punchy after hearing my score so I called 'possible ricochet damage' hoping to put him off his aim. Seconds later he was off target declared ricochets. Time to go home.

The good idea of a mutual inspection on the way revealed long sandblasted furrows and dents under my wings and fuselage. But the puzzle over not slowing down was solved: I had shot my speed brake off and there was red hydraulic fluid seeping everywhere. Dave hadn't faired much better with furrows under his wings and fuselage with red fluid streaks too. He also had an interesting big hole aft of the nose. The fuel tanks were self-sealing but I was a bit concerned about some wetness on one wing. Minutes later I called the tower and told them we had a small problem with hydraulics with possible braking problems. They asked which one. I said both. The long silence that followed seemed full of unspoken questions.

I hoped we would be parked away from prying eyes but the word was out although I must say Wally didn't look as hard-assed as I thought he would. He did raise his eyebrows a bit when we looked at Dave's damage. The 30mm ball round had stopped after buckling the floor upwards. It was directly in line with his bottom ejector seat handle which could have been a bit embarrassing.

Mine was no real drama just a jagged little piece of metal where the speed brake should have been with hydraulic fluid oozing out of a few interesting holes. But an awesome amount of bare metal showed under long furrows in both our wings. Hoping to shift the blame I muttered that the range really ought to be de-leaded after this but Wally just said 'film' and held his hand out.

Coffee was a muted affair. Pretty low-key I would say given the damage we had inflicted on ourselves but when he beckoned us into the cine room I noticed a smile on his face. Not much of one but a smile all the same and considering we both almost shot ourselves down that was a good thing.

We both held our breath as our film whirred around. Neither of us had fouled! 'A whisker away from it,' Wally said, wagging his finger at me and fined us a barrel of beer for the ground crew and grounded us for an afternoon's gardening on the squadron rose bushes.

We, junior pilots, became FNGs – Fairly New Guys and made fully operational as more pilots arrived. Wally and Callum organised the programme around vigorously flown phases which proved an astute move. We carried 100-gallon drops outboard, flew FR navigation phases and practised strikes that took us as far as Norway and France and used firing ranges in the Wash and the East coast. High-level fluid six battle finished by splitting into rat races of three which were high-spirited tail-chases with wingmen reporting the tail end men. These stick search and report exercises were real testers, avoiding the leader was hard but looking back to call on the threatening position of the others under high g turns made it harder still with aircraft all over the sky – utterly brilliant training!

Phases lasted weeks with fluid six battle formations leading to air combat phases where we divided the days up with early one v one's ending up with two versus four. Wally had two Hunters painted with yellow noses to be the baddies for bouncing, it was exhilarating stuff and Wally was in his element with the Gulf and Aden boys. Roger Hyde the most junior vet' had come from a Gulf squadron too so there was a large experience gap between us.

This skills gap in a fighter squadron is well known to all neophytes and the transfer of such skills has been a right of passage for as long as there have been Junior Joes on a squadron. It's about getting to know the new kids on the block as well as preserving the age-old practice of keeping whinging wingmen in place. And it works!

As a result of much practice by we junior joes, the gap was closing with amusing results

chapter Twenty-Three

Dark Arts

AIR COMBAT phases are especially enjoyable and in my case, it was down to discovering the art of crude contest. I noticed fights were often won in the de-brief room which wasn't how I thought it should be and I couldn't help wondering how a confident Wally Willman got the upper hand in every single one-versus-one. I noticed he made a show of taking two film magazines to my one which was always daunting. He constantly plummeted out of the sun instead of arcing up at me where I expected him and it was always my eyes watering from squinting into a white orb. But it was only when he bounced Jerry, Dave and me together that I twigged how he put us on the back foot.

He launched our de-brief with a boisterous: 'Bloody shambles, I got all of you.'

The dishevelled silence said it all.

'OK let's take this from the start – what heading were you on?'

'Jerry?' 'About 190.' 'Dave?' 'Definitely 265.' 'Wyn?' '090 for sure.'

'Wrong, it was North!'

Ashamed at not even knowing which direction we were flying we were easy meat. Then I remembered looking into the sun where I guessed the beggar would come from. It was midday, the sun was to my right shoulder so I *had* been on 090. I wisely shut up and took it on the chin. The way Wally had us apologising before we even started was masterly. I began to notice other things and could write an epistle on the art of underhand methods used. Junior pilots, according to the gospel of fighter jocks, are there just to make up numbers and any constructive remarks by them in briefing should be studiously ignored. Preparation begins when the Leader sidles up to the maintenance guys with false bonhomie and asks what aircraft he has for us...

Squadron aircraft varied tremendously. Some had engines better tuned with a bit more thrust, the airframe and wings didn't have any minor dents at all and their performance was good. On the other hand, some dripped oil, the engine compressor blades were as rough as a hacksaw blade and the engine ought to be changed as soon as you got back. Sometimes airframes were bent, flew a bit sideways and needed to carry trim. Dents on wings from bird strikes and in my case ricochets were as rough as a bear's ass, and guess which ones the junior pilots got?

On a 2v2 sortie, I noticed my aircraft which was a beaut' – newish with my name on it and *just out of maintenance* got snapped up. The leader said it was high on hours and due overhaul so he had better fly it! We, new guys, got the coal-burners – I soaked it up! Briefings were quick with stern looks at wingmen saying we had better know our spin-recovery just to put us off, followed by a few snap questions about an obscure emergency. With experience, I almost mastered every cunning ploy! With a combat phase coming up, junior guys thought me mad asking to be paired against Wally. The art of crude combat normally dictates it's best to take on the Boss because he was always busy. His eyesight is probably fuzzy from the paperwork and he would blunder around out of practice and you could nail him on film. In the debriefing, you would be discrete, diplomatic, and a little coy at claiming all the hits because that way you looked modest and avoided crappy jobs.

The next choice would be the junior shag who had just arrived. He would cop it big time particularly if you put him off his stride when you walk out to your aircraft. Remarks like, 'I'll be watching you so don't break the rules like last time' can be effective particularly if you intend to break the rules yourself.

'Watch your g overstressing is a big sin with the boss matey' easily back-foots him and gets him cautious on the first turn in while you piled on the g to get him.

In between these choices you were forced to rely on skill which wasn't so good – unless you met with someone genuinely hopeless then you stuck with him …

But Wally was not big on paperwork nor was ~~his~~ eyes fuzzy – they were bloody X-Rays! I swear he could see through clouds.

But being a country lad familiar with raptors I decided to do what hawks did – they never moved their heads quickly when hunting unless threatened. Too much ground to cover. They detected movement. And importantly, covered one small area at a time. Give it a go, nothing to lose...

Striding confidently out I ignored his casual banter about taking a few more mags along to film me – the allusions to my last five losses with him I simply dead-panned. His remark to 'watch the g this time' was ironic because I was forever over-stressing with engineers complaining about it.

Blasting off we started at 25,000 feet over the Parish where I checked our heading. No surprises there! If we split to engage I would be looking for his camouflaged jet against the land – difficult.

On the other hand, I would be against a backdrop of clouds over the sea - easy for him! I requested a clearance turn to check the area and cunningly reverse the situation. He told me not to be an ass. I was right!

The next best thing was to do as I was told until we turned in. I deliberately stopped turning after 45 degrees, waited, and then slowly curved back in. Strictly speaking against the rules but I knew where to look which I did slowly and methodically alert for movement. Picking up a black speck moving fast I turned in hard at full throttle...

It ended up one each and a stalemate. I expected an aggressive debrief with remarks about me getting too close and maybe hints about 'cheating bastards not being welcome on a decent squadron'. But it wasn't like that. It was waggish and entertaining because he knew that I knew!

Quick to compliment me after looking at my film he hooted with laughter when I owned up to reshaping a few rules. It was only when I got back to the crew room I realised I hadn't checked his claim but I guessed his film was in the dustbin by now...

Our numbers swelled with Harry Karl, Bernie Scott, Ian Huzzard, Brian McHaffey plus guys from Canberras who were a breath of fresh air. They were all likeable pleasant guys and good flyers. Listening to Howard Lusher over a beer I thought the combination was good and interestingly they all went on to do well.

Life with new chaps arriving coincided with a short-lived low in my life. Chosen as Junior Officer on a Court-martial, I took some Alka Seltzers in the morning to clear a slight headache from a stuffy room. Over the next month, I began getting stomach cramps, lost weight and my flying went downhill. It was difficult to concentrate and I took a bit of stick at debriefs

Matters came to a head when I fired off a burst on the range, pulled off and blacked out. It was only for a second or so and when I landed I felt quite crook. I decided to see the Doc. He was furious and sent me to Ely hospital with a burst duodenal ulcer and minus three pints of blood. A consultant explained it was not unusual for someone who hardly ever took a pill to react this way to aspirin in Alka-Seltzer. He reckoned I was very fit and would be back flying in a few months which was good news. I was pleased it wasn't anything serious.

The boss sent me to a flight safety officer's course. As the new Wing flight safety officer, I was all ears at a meeting about false fire warnings. It was acceptable to ignore a warning if you thought everything looked OK on the dials. Just as you would if you borrowed an aircraft to play golf in Scotland and the system was inadvertently set off by golf clubs falling on the control box. Trying to impress I added my pennyworth that the system was a Fault Free Fire Detection so no such thing as a false fire warning – technically speaking.

'Right we have no further false warnings chaps,' declared our new Wing Boss Sharkey, 'From now on they are all real and bloody dangerous and if the lights are still on you bang out.' And that was that.

Days later I looked up as two Hunters broke into the circuit with their fabulous 'blue note' and was surprised to see one plunging vertically into the ground.

Turned out that one of our fairly new guys had a fire warning, closed the throttle did the drill waited for the light to go out and when it didn't, being a conscientious chap he remembered the Boss's statement and promptly and rightly banged out. So everything was fine.

At the hastily called safety meeting, I nodded my head to acknowledge we revert to the old ways with a fire warning light. Except for golf club and squash racquet-induced warnings, we would try for a visual check and play it by ear before banging out.

Returning to the skies I had to get up to scratch again. With no stomach pain, problems or puking, life was a breeze. I no longer felt listless but confident and robust.

As a low-wing loaded aircraft, we were good for combat with other types such as the Lightning and Phantom and the odd Mirage. The tactical view was that they hadn't made a head-on kill and closed in for aerial combat.

And I learned many things.

If you took on a Lightning above 25,000 feet – *You* lost!

If he took you on below 25,000 – *He* lost.

A Phantom, with a macho crew, fully Brylcreamed with many missiles could plug in the burners to show you who was top-dog and pull a lot of g.

But if he turned to smirk at you with a pair of blazing afterburners and pile on g to get behind you – he stayed in front. You simply pulled inside at 450 or so at full chat. He hurtled around a bigger radius at Mach-something used all his fuel and stayed in gun range for ages. Great fun and we got quite good.

The most amusing was the least mentioned. The Harrier bona mob had a go at us.

Being a humble Hunter pilot flying an obsolete jet that didn't do vertical things, I was asked how we expected to manage a fight with such a feeble engine. Get real man. Get a Bona-jet! They offered to run a book on fighter affiliation with us and it's such a shame we didn't. Having been told 'VIFFING' was the in thing – I asked what the hell it was? – 'Wait and see.' Variable In-Flight Forward thrust would confuse the hell out of us. It was the equivalent of putting on a hand brake in flight, and it certainly confused everyone, but not us. A notch of flap a high-speed barrel roll a dab of speed brake and it was like shooting fish in a barrel. 15 bona-jets to one Hunter.

I think they did something highly technical like beefing up the bicycle chain used to vary the nozzle angle. But it was obvious the aircraft had qualities no other jet possessed and would change aviation. A few of us were asked to convert. Bernie Scott did and I know he never regretted it, I was tempted but was enjoying myself while I could on Hunters.

I was polishing the dark arts!

Some pilots get worked up on 'one v ones' becoming highly aggressive at times. The results are often hilarious with a few tantrums! Payback time is healthy and good for any squadron because it means the system of training is right, guys are improving and everyone benefits. It's the only thing in aviation where you can prove yourself an ace by constantly being behind everyone else. Life for us junior Joe's boiled down to beating our early mentors with style.

While innocently minding my own business Roger strolled into the crew room with a cheeky grin. He said two aircraft had come up, he was authorised for one v one but he couldn't find anyone good so it would have to be me. I bounced up heading for my kit. Brilliant. He had started nettling already! Must be worried. It was going to be a case of revenge is sweet best savoured cold in gun range.

Roger was reckoned to be good so I knew I had to recall everything I had learned. But I didn't mean honest fair stuff obeying the rules. No, it was time for the art of crude combat where the best cheat wins! – Roger was too honest for that you see. I made use of the brief to wind him up: I had been shown by experts how to do this and when asked to quote the rules I made a show of counting my extra cine magazines. I had to be asked twice with growing irritation. With a face like Attila the Hun I asked if we were sticking to frigging rules or what? I couldn't possibly remember all of them but who the hell cares as long as you don't hit each other; I mean this *is* a fighter squadron, not a *bloody girl's* school. You could almost see the steam coming out of his ears by the time we signed out.

We hurtled off as a pair with me on the wing at 25,000 over the Parish. I didn't like the heading which would have disadvantaged me, I knew it wasn't deliberate on Roger's part he was far too rule-abiding and nowhere near as dishonest as me so I called a false contact. We turned to engage, I said it went into a cloud and suggested a heading to keep clear…

I think the low point on debriefing was when I offered to swop aircraft for another go just to make sure he was OK on power this time! In the end, he saw the funny side and we had a few fun beers afterwards. And do you know what? When I said I hadn't bent the rules much on my first claim but the other two were kosher he had the cheek to say he didn't believe me. Who cared! Getting to the top of the one v one ladder was swagger stuff. A high-level detail against Cyd Sowler ended in a stalemate which pleased me no end as he was brilliant and during our debrief, Boss stuck his head around the door and asked if I could take a Bucc slot in a couple of weeks. I stuck a thumb up and that was that!

For my last range trip, I was given rockets and full guns. I saved the guns till last and decided to switch to 'all' instead of pairs. The airframe vibrated, four cannons hammered beneath my feet, circuit breakers popped and fuel booster lights came on. After shoving all the circuit breakers back everything worked and my best score ever! I even remembered to square it with the plumbers when I landed with no ammo left for the next guy. A great way to finish on Hunters, I learned much nobody had been killed and we only lost two aircraft, one to a fire warning light and the other in bad weather after takeoff.

Before flying Buccaneers I had to visit RAF North Luffenham in Rutland to be measured for a personal harness and decompression runs in the chamber where we simulated ejecting at 50,000. Rutland was a small county with a big reputation for brewing the famous 'Ruddles County Ale'. A nutty beer but gassy, the importance of which shouldn't be overlooked because Ruddles County Ale enjoyed a symbiotic relationship with the aero-medical doctors. The mutual benefits became obvious after a rapid decompression. Docs emphasised that in a sudden loss of pressure, the rapidly expanding gas in miles of our intestines and various cavities had to come out. Healthy belches and farts are required. Ruddles Ale not only assisted the process but provided ample evidence that you had done so. The odour in the chamber was appalling.

Years later I related these experiences when training civilian pilots of high-flying business jets operating above fifty thousand feet. Not one of them had ever done a decompression run, which says a lot about the thoroughness of our RAF training. It always surprised me how a thoughtful silence descended on my lectures when I mentioned the insidious effect of short-term memory loss that comes with hypoxia. And if you were lucky to survive a hypoxic situation and land OK, it would have been the best trip ever and nothing wrong with the aircraft! The next pilot flying it might not be as lucky...

chapter Twenty-Four

Buccaneers

image James Biggadike

IN 1974, I pitched up at RAF Honington to fly an aircraft the RAF didn't want to buy. Now they couldn't get enough of them and because of the enormous fun to be had flying the Buccaneer S2 every pilot and his dog wanted to fly one. For this, we have to thank the Navy because they got it first and ironed out the wrinkles.

RAF Honington near Bury St Edmunds in Suffolk was home to the Buccaneer Operational Conversion Unit (OCU) and 12 Squadron the first RAF Buccaneer squadron. After converting I was going to 208 Squadron which was reforming, an arrangement I was completely happy with because it was a sought-after job not open to first tour pilots.

Two other Buccaneer squadrons: 15 and 16 Squadrons were in Germany where they had already gained a strong reputation for low-level flying compared to other air forces.

Training at the OCU meant re-acquainting with the Royal Navy. Previous conversions had been undertaken by the Navy but with the scrapping of all but one heavy carrier, the number of RN pilots declined so the RAF had taken over the job but retained a small number of experienced Navy guys. There was nothing new in the RAF bolstering Fleet Air Arm squadrons, they had done it for many years on many carriers. But I knew it could be a sore issue, I just hoped they were unlike the surface chaps who held definite views about us Crabs. For those unfamiliar,' crab', is a polite insult to make sure the RAF junior service knew their place in life. I liked the name. Over a welcome barrel to my great delight I met Ian McKenzie, he had taught me to fly on JP's and having flown the Lightning was now on the same Buccaneer course. There were four pilots, the other two were Trevor Brown and Royal Naval Lieutenant Perry. Our four navigators, Barry Southwould, Phil Ford, Jim Crowley and a first tourist Pilot Officer Craig held beers with smiles that suggested they were happy to be where they were – a good start.

Feeling upbeat, I hailed Perry with a jovial hello who gave me a withering look and informed me in tones of righteous indignation that the Navy should be doing his training and not the bloody Crabs! This rather dashed my hopes, it was clear that aversion to the RAF was a whole of society thing with them so I had new etiquette to navigate here. Not one to make a fuss on the first day I spoke slowly as if to a foreigner with questionable wits. 'And what Navy would you be from then?'

I think it was Ian who intervened before my diplomacy had run the full course.

We Crabs and our new Navy friend actually got on well and everyone on the OCU was good company. Graham Smart was the Chief Instructor. A burly-looking chap with a shock of dark hair he had a pleasantly tough manner about him and every instructor was upbeat about the Bucc', and I found the dark blue embraced flying in much the same way as the light blue.

This was the second Blackburn aircraft in my life and I hoped the design team had changed from the one that decided the engines for the Beverley.

My worries were needless. The Bucc's Spey engines didn't need an oil bowser to top up after a long taxi and technical systems were easy with a practical electrical system. The emergency battery power was impressive at about 80 minutes and it was interesting the way Naval matters influenced the design - wing folding being obvious. The hydraulics were practical, insofar as you needed to get the gear up quickly to avoid drag after a catapult shot so the main hydraulics worked on 4000 psi and flying controls left to a more standard 3200 psi. After a leak the practical seals allowed the reservoir to be filled with coffee or such to get you back on board.

'Would that be percolated or instant coffee?' I asked

'Percolated Colombian Best for the Navy naturally and cheap instant crap for Crabs!' Doesn't pay to be smarty pants with a Navy instructor!

Viewed from the front the Bucc looks purposeful with striking high-speed looks but from the side, it's as ugly as sin. The reason for the balance between beauty and downright ugliness was all down to 'Area Rule' where the cross-sectional area of high-speed aircraft formed a smooth streamlined curve from nose to tail. With Area Rule, the area of the fuselage in line with the wing was reduced and the area aft of the wing increased.

This resulted in a big reduction in drag which meant less thrust needed at maximum cruise speed. The design was aimed at supersonic flight but the Blackburn boys adopted it and gave the Buccaneer a wasp waist and a bulge in the rear fuselage. This delayed the drag rise at speeds close to Mach 1.0 improving the flying qualities at high speeds and the result was the finest low-flying aircraft ever built.

The 'back-end bulge' was good for packing in equipment (I once stowed several hundred-weight of Norwegian spruce to build a Jacuzzi with plenty of room left for golf clubs and fishing rods)

Buccaneers are a potent weapons platform due to their rock-steadiness at high speed. It carries a formidable amount of armament outside on pylons and inside on a weapons bay that rotated open at high-speed at any height. Built for the Navy it had certain limitations for striking below deck. The Navy is not into raising hangar roofs or knocking out bulkheads so the Buccs length: some 65 feet was shortened by a folding nose and the airbrake petals which stretched to the end of the fuselage opened fully. Folding the wings saved almost 20 feet but the fin had to be shorter than required aerodynamically.

What you didn't need was a wing drop with one wing stalling before the other while attempting to fly as slow as possible to hook on. The option was either to use slats or leading-edge flaps on the outer part of the wing – or try something new. Brough decided to use BLC - Boundary Layer Control – using engine air blown through slits in the leading edges of the wing instead of leading-edge devices. Blackburn never did things by half, they blew engine air through slits over the wings, tail-plane and the wing trailing edge.

By drooping the ailerons you effectively ended up with full-span flaps and this reduced the landing speed by a minimum of 17 knots which was good news for hooking on. This BLC provided a solution to anti-icing but it wasn't the complete 'answer to the maiden's prayer' because it used a lot of engine power and invited potentially dangerous pitch changes. So Blackburn designed tail-plane flaps to overcome the problem.

The position of the flaps ailerons and tailplane flaps are forever inscribed on the crew's memory: 45-25-25 blown and so on!

Blackburn decided to put 'zing' into the Bucc with the more powerful Spey engine which gave plenty of blown air and provided excess thrust. The Buccaneer S2 could motor and as a consequence, could easily exceed the maximum design speed of the airframe in level flight and just as importantly the limiting speed for weapons carried on the pylons.

As a two-seater, we paired off and I imagined I was the last person any navigator in his right mind would choose to fly with. I imagined the process would be like school when two ace footballers chose teams and you were the last one remaining. But it wasn't like that! And I don't think they used a cattle prod on Barry either and as the course progressed I realised I was damned lucky to fly with him.

Buccaneer crew terminology was straight out of 'Jeeves'.

The term pilot or navigator is seldom used. He was simply – your man, which works both ways. Either that or front seater or back-seater. The philosophy of a two-man crew has a powerful edge to it which I promptly grasped so it was no good playing the ex-single-seat–who – needs a navigator – card!

No. I had been earwigging at the bar picking up crumbs from the Fleet Air Arm who casually mentioned flying at fifty feet in crap weather doing 500 knots over the sea was a piece of cake. You just needed a burst of radar now and again, even a Crab could do it! I think they probably clocked me hovering nearby pretending to be disinterested.

To fly this beast we needed full support from each other.

And that wasn't counting engine fires and other big emergencies plus getting me out of the kind of trouble I seemed to have no problem getting into.

chapter Twenty-Five

Rock-Solid-Ride

BLACKBURN WAS frugal not building a dual-control Buccaneer which added to the excitement of flying one for the first time. Clever thinking by the Navy resulted in dual Hunter T7As and T8Bs fitted with the Buccaneer IFIS – Integrated Flight Instrument System. It was a smart move that gave pilots a good feel of instruments and thanks to the Navy and Blackburn I kept my hand in on the Hunter. Couldn't get any better.

On the 24th of July '74, I flew with Ian Ross an ex-DFGA Gulf Hunter pilot who was good news. Ian was a first-rate pilot and a friendly guy who gave me advice on Bucc handling and the hot poop on life with a fast-jet navigator.

On the morning of August 4th, I walked towards Buccaneer XN 976 with Hilton Moses to begin the external checks. The cockpit was agreeably larger than a Hunter, the ejector seat was as comfortable as an armchair, rocket-equipped it gave a zero-zero capability – zero feet zero airspeeds. A great office.

Hilton was an experienced pilot who frankly deserved a medal every time he flew with a brand new Bucc pilot. Contrary to popular folklore he wasn't fast-talking with no stutters and neither did he fly with one hand on the 'bang' handle at all times. He was calm and laid back.

Any pilot changing aircraft goes through a subtle period of skill sensitivity in the first few sorties. The previous casual familiarity with checks and systems is now elusive and any awkwardness in the cockpit can make some characters touchy and permanently affect their feelings towards their new work-horse.

Hilton showed just the right amount of consideration towards me stumbling through the drills. Not easy for him to balance my 'let's get this beast on the road attitude' with my excessive zeal for a new job with zero experience in it.

For the occasion I wore my new green full torso harness; a comfortable and easy plug-in for the g pants compressed air, oxygen and communications and in the back, Hilton's view forward was good because my seat was offset a few inches to the left and his to the right. The Spey is started by Palouste air and being a twin-spool fan has a comforting growl and airframe shudder on starting. And away we go! No fuss.

Take-off is unique. With no dual, it's an adventure into the unknown in a pokey performance aeroplane. The nose-wheel steering gave a super accurate line-up and acceleration good with a satisfying push up the rear and a short take-off run.

'Nine Seven Airborne to Zone.'

'Roger Nine Seven Stud three'

Clean up's rapid with beefy hydraulics. The wheels clunked up, selecting flaps up I hauled the nose around towards the coast keeping an eye on my left upper side instrument panel 'cheeses'. These monitored the position of the flaps, ailerons and tail-plane flaps and they needed to move together, if they didn't it was bad news, a rapid pitch problem and a hole in the ground if you don't stop them. Apart from this minor idiosyncrasy, it was easy to forget you were in a powerful aircraft.

Immediately after we cleaned up, the handling characteristics showed up. A comfortable sturdiness to the handling without the controls feeling heavy. Accelerating in the climb through 2000 feet I was aware of the smoothness of the ride and the absence of the lively bounce of the FGA 9 in turbulence.

Hilton talked me through the handling which was a matter of steep turns barrel rolls and general handling and then a climb to 30,000 to get the feel of power and airbrakes up there. The latter was controlled by a rocker switch on the inboard throttle, the huge airbrake petals were quick to extend and even quicker to close. Flying at 540 knots slamming throttles shut and flicking the airbrakes out threw you forward into your straps. Impressive stuff and good news, speed agility is a nice attribute in air-to-air combat.

Power off. Brakes out. The rate of descent was over 16000 feet per minute plus and looking sideways we seemed to be at about 80 degrees to the horizon going straight down.

Very easy to go supersonic (not allowed) or exceed max indicated airspeed. Slowing down to 300 knots at 2000 feet I powered up against the airbrakes and then simultaneously went to full throttle and selected them in. With no change in handling in less than 20 seconds, we were at 580 knots diving toward the sea. Levelling at 200 feet the aircraft felt rock-solid. *A dream to fly*. Couldn't wait to fire off weapons.

I took it lower. As an upshot of area rule aerodynamics, we were not battering our way through the air; more a sense of the sky cleaving apart for us. Utterly brilliant. Time to go home and do some circuit bashing. Asymmetric work was new but the yaw was nowhere near as fierce as I thought because opening the speed brakes placed the petals partly in line with the engines so the good engine efflux offsets the yaw.

My next trip was with a fully paid-up OCU back-seater. And I couldn't help wondering how it would go. Bound to be fun.

The OCU lads ordered me a packed lunch along with a few navy guys too who weren't hopeful of the outcome but having survived worse in Coastal and not being riddled with food anxieties like the Navy, I tucked into hard bully beef and curled up bread with relish. A simple brief by Lieutenant Ritchie for my first fam' flight. 'A good throw around getting a feel for the aircraft at low level' is exactly what he said – and I intended to take him at his word. He had mentioned tailplane blanking in a loop and no-no's like spinning but I was thinking more about the low-level throw-around, to be honest.

An old photo of a 'Vic' of three Buccs going over the top of a loop caught my eye, it was a dynamic photograph somehow at odds with my brief.

Out on the line: observing my ritual for soloing a new type – I kicked the nose-wheel shook my fist and gave it the 'remember I'm the boss' routine before strapping in watched by a gob-smacked ground crew.

On my second flight, I was amazed at the things I missed on my first. But I hadn't missed the audio ADD: it stands for Airstream Direction Detector and is a simple and effective angle of attack designator and a great safety feature. Designed for carrier approaches we used it on the first trip for circuit work to get the optimum speed. A steady sweet note meant speed was smack on. Just a few knots slower brought in a low-toned insistent burp, a few knots faster and the note became higher-pitched beeps. The beeps or burps could blend with the sweet steady note as speed increases or decreases through the margins. My man suggested I switch it on for general handling.

Coasting in and out points on the Norfolk coast helps ATC to pick us up and de-conflict from Lightning and other traffic. In minutes we were there with the blue-grey spirit-level of the North Sea filling my windscreen. I tried a few aileron rolls to loosen up revelling in the fast rate of roll for such a big aircraft; the powerful ailerons were large (these were geared and when the speed dropped below 300 knots pulling up a handle enabled different gearing for lower speeds) I tried a dab of the rudder and the roll was crisper. Ritchie was full of helpful chat and I was confident nothing could go wrong on a fam' one flight.

Cruising along nicely over the sea I recalled the photo shot of looping Buccaneers which promptly brought with it 'anything the Navy can do I can do better' feeling. Should have ignored it but memories of Dai Heather–Hayes and Wally Willman's fighting culture were too deeply ingrained.

It did occur to me that the first thing I was doing was the very last thing I was told not to do but I was at a handy height anyway and doing 500 so I pulled up with 5 g and powered upwards. The altimeter promptly passed 5000 feet I kept the pull going and heard a small groan from the back, the audio ADD worked a treat and over we went!

The Bucc was stable and controllable with no tendency to tuck in, easy to keep the wings level and I made sure we had plenty of speed with no possibility whatsoever of blanking the tail. Easing the back pressure off at the top I sensed the potential energy as soon as we headed down. My man started briskly calling heights and speeds which impressed me so I took the hint throttling back gently which turned out to be a good thing as the speed builds up rapidly. Had to watch that! I kept the pull on to get some g on to slow me down but was still ready for a touch of airbrake. God! I already loved this beast.

I went straight into the next loop with Ritchie calling speeds and heights, bottoming with 480 knots I edged down to 100 feet and found the Navy had got the low flying bits well-sorted upfront.

Big precision ASI close to my left eye for deck landings. An accurate radio altimeter with scales for 0 to 5000 feet or 0 to 500 and variable height traffic lights settable to different heights. These were highly visible next to my right eye, I set this to 100 feet and now had a green when at selected height, amber slightly high and red slightly low. I forgot what the 'slightly' was, I think it was 3 feet or 3%. Not that it mattered, if you got a red you did something about it!

Flying at 100 feet 480K was a piece of cake, the handling was superb and banking in turns the nose held up effortlessly with a slight pull. Visibility from the cockpit was brilliant; turns with a spirit level sea across the big windscreen like a giant artificial horizon made it easy.

I got tons of advice from the back. At 420 knots at 250 feet the manual quotes 100 lbs a minute fuel burn and 580 knots, the burn was 120 a minute. Not a huge difference.

I shoved on full power felt a lovely surge and quickly took it off as we hit 540 knots. The ride was rock-solid and being a fighter disciple I tried swivelling my head around for a lookout. Having the confidence to do this in a new jet impressed me. Over the sea I wasn't going to hit anything; unlike flying overland but even so, I had developed a tendency to restrict my view forward in the Hunter at the same height. A big difference!

Speed was restricted to 580 knots by the OCU and I throttled back to maintain it. Conscious that with tail-plane height restricted by design the directional stability might not be quite as good in a turn I tried to be smooth. Good ride. Although I felt a boot of rudder at that speed would be unwelcome, maybe push towards the grim possibility of inertia-roll coupling. It was early days on the Bucc so best not to be overconfident.

500 knots plus felt completely comfortable so with 50 feet selected on the rad alt and my man happy I eased down and cruised along with the waves flashing past giving a terrific sensation of speed.

Had to watch the turns being conscious of wingspan and to be honest just that difference of 50 feet concentrated my view forward. I slowed to 480 knots and felt comfortable swivelling my head around immediately. It pointed to a lot of practice flying on the wing down here, especially if you wanted to manoeuvre hard. The Fleet Air Arm boys said they did it all the time so I knew where the bar was set for us Crabs!

All too soon it was time to go home.

Breaking into the circuit and changing gear on the ailerons changes the thought process towards different handling at a slower speed. The landing wasn't particularly difficult as it was designed to be flown straight into a deck with no flare. I felt pleased. My first trip flying a superb aircraft with an enthusiastic guy in the back with shed-loads of experience. Brilliant. What's more, I was surprisingly unfatigued – but ready for a coffee. My instructor disappeared to the instructor's room to mark my file and I settled back to wait under the photo I had glimpsed before taking off.

On closer inspection, the Buccaneer formation in the photo wasn't looping at all. They were in a steep barrel roll: the blurred horizon had fooled me. Oops!

chapter Twenty-Six

Consequences

image © Michael Rondot collectair.co.uk

WITH TWO and a half hours of Buccaneer time under my belt, I looked forward to flying with Danny Daniels a Naval Observer. A jovial character, Danny was highly experienced flying off carriers and I assumed he would encourage me to throw the aircraft around. And I would. I decided my temperament would be better suited to the Bucc if I adopted a forward-going attitude as you would with a big horse. No point in shying back and saving the spurs otherwise you wouldn't get the best out of it.

The weather was perfect but a small problem when I signed out – someone had pinched the radar. They had bolted it into another aircraft and left me with lead ballast and a note in the F700 – maintenance log – that said the handling properties might be marginally different. Not as nose heavy so I remarked to Danny I would check handling with a few gentle aero's before we got into the low-level stuff. He showed no concern and on rotation, I noticed the nose came up quicker during the rotation but otherwise it felt OK. Safe as houses.

At 5000 feet and 500 knots, I called 'starting aero's and tried a few snappy aileron rolls and it felt fine but pulling up to the vertical I noticed the aircraft tended to tuck in and I decided to see what it was like over the top. Danny had been offering good advice and as I was now totally into remarking what I was doing next: I called 'looping' and was mildly surprised to hear *'bloody hell you ARE going to loop.'*

There was a different feel at the top and I could feel the aircraft wanting to tuck compared to my previous flight so I said it would be unwise to do it again. I think I detected relief in his generous thank f**k for that! It made me feel good. Picking up points for doing something responsible like not risking the destruction of a Buccaneer on your second familiarisation trip was bound to be useful. It was reputed to be a tough course so I needed a few points.

Danny gave me excellent low-flying advice on the 100 and 50-foot stuff even though we should have been at 200 feet for most of the sortie and I was struck by the way he calmly phrased everything. We ended up well north of the Wash area and turned for home, feeling happy and confident with the handling I suggested we close with the coast to pick up some fast traffic. I filled in a momentary silence by saying I used to bounce stuff there; his laugh confirmed he either thought I was joking or he agreed. I settled for agreed.

Climbing to 2000 feet I spotted a US Phantom and made for it. It was a bit of a turning scrap and I was impressed by the way the Bucc handled it: we were not being waxed at all by a fighter, far from it. With the lead ballast in the nose, I noticed a distinct tendency again to 'tuck-in' with g but as I was attempting a 'rolling scissors' at the time to get behind the F4 it helped. I was having a whale of a time but sensed I had a lot to learn. At this point, I must say, with powerful airbrakes and high thrust the encounter demonstrated to me the capabilities of the Bucc more than a score of briefings could. This aircraft was full of surprises and although I knew the OCU pilots would have done better it showed me the potential and that's what I presumed these first famil' sorties were all about.

Walking towards the crew room I noticed Danny grinning a bit tightly at my enthusiastic chatter and I wondered if I had screwed up over something. He disappeared upstairs and said he was going to see the CI for a few minutes so I made the coffees. Looking around at grinning faces with matted hair slurping coffee and gesturing with their hands; I guessed the others on the course were having a good time too. Wandering over to Trevor and Ian I asked if they had fun mixing it with anyone yet? When both shook their heads I remembered Danny's remarks and had a sinking feeling that a whole banquet of consequences might be coming my way.

But Danny chatted about our trip with the upbeat natural enthusiasm I had come to expect from the Navy guys; although he did finish by saying he felt I would enjoy the course and would I mind stepping into the Chief Instructors' office before I went home.

I knocked politely and entered to a cheery 'come in' to find Graham Smart tucking his flying suit into the top of his boots. 'What do you think of the Bucc then Wyn?' he asked in a friendly tone. I launched into a 'bloody marvellous super ride down low and love the power…' He grinned at my dissertation and asked:

'Any criticisms?'

'Could do with some cannon' I promptly replied 'then it's got everything.'

I gave a brief account of bouncing the F4 to back this up. As CI I thought he might be interested in first impressions of a happy neophyte joining the force and knowing Danny had seen him before me: honesty was the best policy.

He looked at me thoughtfully as if he was making a decision; gave me a nod.

'Tell you what Wyn after you finish the course let me know if you changed your mind about that' he stuck his hand out 'Good luck.'

On the way out I began to wonder if I had overdone things. But Hell! This was a fast jet and that's what you do. Nobody told me you shouldn't bounce anyone and it's exactly what Dai H.

Hayes and anyone of spirit on Hunters would have done so I put it firmly out of my mind.

Graham Smarts' manner impressed me. He accepted my enthusiastic handling was genuinely inquisitive rather than unimpressively cavalier: which would need censoring leading me to the opinion that flying on any squadron he commanded would be good news.

It was a reasonable start and anxious to keep a low profile I forgot about it. Besides Barry and I, or Baz, as I sometimes called him were going to fly for the first time as a crew, something to look forward to. Together we would fly set exercises, sharpen our cutlasses on the ranges and hopefully bounce a few instructors to see what they could do. I was bound to learn something there. Air-to-air refuelling would come later on 208 where our role included nuclear strike; which meant nuclear indoctrination to carry the neat WE177 weapon.

Every time we flew a new phase an OCU backseater would fly with me and Barry likewise with an OCU front seater then we pressed on as a crew. It was a steep learning curve with both of us determined to make it work with Barry proving to be as sharp as a pin on fast jets in a whole new dynamic environment. I always tried to warn him of rapid manoeuvres when I threw the aircraft about which hopefully helped and as for myself – I never had it so good! I always carried a map but now I get to sit in a bang seat as comfortable as an armchair while Barry sorted out where I was! A dream job.

With him doing all the work I had more opportunities to look for trouble so I stopped advising him where I *thought* we were because he always knew *exactly* where we were. That was neat.

Formation flying was interesting. Standard wingman positions were easy but I noticed flying in formation with flaps down on a 'blown' approach there was a tendency to be sucked in if you got too close – something to be careful about. Where the Bucc also shone was on tactical join-ups with its speed agility: joining up quickly with bad weather looming was easier than the Hunter. A big dab of power got you moving in fast and the mighty airbrakes killed a big overtake quickly and bingo! You were on board. A handling quality scarcely mentioned.

chapter Twenty-Seven

Sharpening The Steel

I LOOKED forward to weapons training as much as I looked forward to Christmas as a boy. Baz had his Blue Parrot Radar to play with too. Most of the OCU back seaters said the Blue Parrot radar was useless overland but that was their chat, not mine – I was grateful for any kind of kit. It might be analogue stuff but there again so was the Harrier's inertial kit and the bona-jet boys, bless them, managed to hit most things they aimed at – so they claimed anyway!

Buccaneer attack kit might be timeworn but it didn't constrain my optimism, the accuracy of automatic release attacks wasn't shabby at all. Bloody deadly was more like it. Built to sink the big heavily armed Sverdlov class cruiser pride of the Soviet Navy it could also take on and destroy fast patrol craft and sink the largest merchant ships both day and night. I had seen more than one Sverdlov close-up plastered with guns and dangerous-looking so it followed you needed mean weapons to sink them. A nuke is mean and what's more the Bucc could carry a brace of them.

To Soviet eyes, a Buccaneer was a deadly threat so they sensibly stopped building Sverdlovs. Forty were planned but construction stopped at fourteen with the Royal Navy justifiably claiming carrier-launched Buccaneers helped the Kremlin with their decision-making!

Bombs and torpedoes gave way to missiles and although the bombing option was retained, the first conventional choice was now a missile. Martel missiles, TV guided with an anti-radar version – ARAM – were weapons of choice and the Sea Eagle was coming off the drawing board. Fascinated by the amount of damage you could cause with just a small formation of Buccs I knew with certainty I had a calling in life!

A typical four-ship attack formation would have three aircraft equipped with three TV-guided Martell missiles plus a video pod each to enable the back seaters to guide them and a separate load for one aircraft – usually the Leader – with four anti-radar Martells (ARAMs) which were very accurate.

In addition, each aircraft would carry four 1000 lb bombs internally on their rotating weapons bay, which could be used if the weather prevented TV Martell homing.

For the pure bombing role, underwing stations carried four 1000lb bombs plus four in the weapons bay. Bombs – including nukes – could have tail parachutes to retard which prevented the bomb from exploding under you, or ballistic for tossing. Fuses were impact, delayed or VT variable time which exploded at 60 feet.

Fast patrol or missile-firing high-speed boats can be a real nuisance at sea. They were a cheap option for some nations and a Bucc was the complete off-the-shelf answer to them; in the day no matter what speed they were doing you had them cold. At night Lepus flares of a million candle-power tossed from a few miles away illuminated the boat and the same aircraft armed with pods of two-inch rockets would re-attack the boat silhouetted against the flare.

We had two ways of aiming at a target – visually or by radar.

Level laydown attack between 150 and 200 feet for a visual attack or 500 feet using radar. Dive-bombing speaks for itself; the dive angle varied but they were all visual.

Toss attacks involved pulling up smartly and the Control and Release computer released the weapon. A medium toss for iron bombs and a long toss for nuclear bombs – vari-toss generally used on overland targets was a timed attack and the speeds for attacks were fast between 500 and 550 knots.

After weapons famil' Baz and I filled our boots with range work where I found the Buccaneer a hugely stable weapons platform; a quick dab of the rudder on rollout stiffened the dive and a quick 'ball-in-centre' check with two engines paid dividends and scores improved.

Next up Strike Progression – STRIPRO – flying in low-level battle formation to a target and being bounced. A Naval term for something we did routinely on Hunters so I ambled into the briefing room all fired up for a scrap.

Evading fighters to hit the target is common sense and sometimes means forcing them to fly opposite to your flight path going in the wrong direction. The big issue here is that you are inviting the beggar into your 6 o'clock and because he was bound to chase, our 'running out' speed was vital; you had to destroy a target not pussy-foot playing the fighter ace. So 'Evade' on the board in big letters was the aim today.

And it was just an innocent remark of mine that caused all the trouble.

The OCU Lead glanced at me to emphasise the big word. Given the three-position, I assumed I was deputy leader and being squeaky new to Bucc tactics asked if he would like a wide battle for missiles?. Wide battle makes an attacker commit to one target while we haired off and I didn't regret asking because I got a thumb-up. But I did regret jesting we had bugger all guns to fight back anyway because a youngish OCU back-seater thought it heresy for me, a new boy, to mention the 'guns word'. He got quite stroppy and misled me by saying the Bucc wasn't a pushover because they carried sidewinders.

Nobody said you couldn't simulate using it! Moreover, I swore I heard the Fleet Air Arm boys in the bar say they could carry a 'winder clipped to the same pylon as a bomb'

I may well have been wrong of course. So I wisely shut up.

The visual lookout was brilliant with two pairs of eyes looking for threats now so I expected a belting good scrap. And it was! But my calls weren't good.

I messed up big time.

'Counter left or Counter port or starboard go!' meant to the OCU a hard 90 degrees or so turn to the left or right. But it confused me when somebody made a counter call failing to back it up with a good heading call as you would if the bugger was right in amongst us.

Sighting a Hunter arcing into position in the leads six o clock; my call: 'Break left go! Roll out heading 250 and I'll sidewinder the bastard' didn't go down well at all. Nor did the second. Trouble was I felt at home down in the weeds and to be honest I tried to keep quiet but it's difficult after being psyched up to be highly aggressive for years. So I did a split-ass turn and hacked the bounce. Our course had hardly begun – how cool was that?

Wrong! Not cool at all. And as is the way of two-man crews, twice as many people joined the queue to tell me off for my unseemly behaviour. With all the verbal flak I twigged immediately that the real message was that a few staff got caught with their knickers down by Barry and myself. I kept quiet.

My errant sin wasn't a show-stopper. No way. This bawling out was chivalrous compared to the rich hostility of my previous world of near-collisions and accusations of rule-bending and gross unkind remarks about your flying.

Anyway, the Lead seemed to like the way I got stuck in, so I felt happy and as a raw beginner on the force it was a good introduction to the ethos of the Bucc and the way the OCU crews taught tactics.

After familiarising myself with the strike sight we dive-bombed our way around the ranges and completed a progress check. I flew with Dave Wilby an OCU back seater for visual level-laydown attacks simulating retarded weapons. I liked Dave straight off. He was a stocky bouncy character with the cheerful look of a fit flyhalf about to trample on someone and we got on famously as soon as we fell out on how to aim the sight.

A simple misunderstanding. The brief was to be level at 150 feet and bang on 500 knots with my sight running through the target. How we got there I assumed was up to me. He thought I would descend to 150 from higher up. But Wainfleet Range was nice and flat so I hammered in low and *eased up* to 150 feet making sure I was nice and settled to pickle off (press bomb release). It was clear Dave understood me and being a punchy instructor suggested we try running in from various heights emphasising being steady at correct height release. No nit-picking – brilliant!

But I was less than happy when we started night flying.

Being an ex- day fighter soul '*night*' to me meant being tucked up nice and safe in the bar when it was dark and scary outside.

There were inducements to fly well in the dark, attacking at night was one and the second was night air to air refuelling where night formation needed to be up to scratch just to get to the tanker never mind prodding away at a hose basket in the dark.

Sweaty stuff I was told and we would do it a lot on 208 Squadron and what's more, they were allegedly planning to do it tactically with no lights on the tanker. Gritty.

No pussyfooting with a Hunter dual night check, it was straight in and I wasn't unspeakably scared like beyond words scared. Well, maybe a bit to start with but the radio altimeter made it a piece of cake over the sea so I was happy. Right up until we started night close-formation and I hadn't foolishly listened to scary stories of inky black nights, clammy hands with eyeballs on stalks peering into a dark void. Being told that you were either out of your mind or had to be ten feet tall and fly like a god if you broke off and tried re-joining onto a dark shape hurtling low over the sea.

And guess what? The stories were completely understated and it proved even more bloody terrifying! Nights were pulling in which meant earlier finishes to night flying and more time in the bar. As we were into the shooting season I supplied game birds for some private course dinners, the rest of our course bought the wine, the mess chefs produced fine dining and as everyone was doing well the feeling was upbeat. November was to bring more challenges with night weapons. But sadly for the Fleet Air Arm guys, the month brought tragedy.

chapter Twenty-Eight

'Tragicum Nocte'

WINTER NIGHTS darkened our November Suffolk skies early. Having fewer night hours than the others called for harder concentration on instruments to back up my visual take on everything. On fast run-ins over mudflats of coastal ranges, there weren't any objects of sufficient bulk to provide scale to the tidal pools and shadowy mud creeks. Visual height perspective subtly altered with the passage of clouds and tidal pools could be larger or smaller than you thought which in turn put you higher or lower than you were.

Briefings are pithy and to the point.

'It's difficult to judge your height in all that crap guys so be bloody careful and watch your rad alt like a hawk.'

But radio altimeters can fail!

Dark shadowy pool-strewn shoreline reaching to wintry horizons blurred with mist and ghostly fog might be an artist's dream to paint, but, it wasn't a joyful prospect to anyone attacking the rusting hulk of a ship target half sunk in the mud. A high-speed shallow unintended descent was difficult to pick up and a sharp turn with no horizon was bad news.

Attacking at dusk could end up a real nightmare because it was visual aiming and at 150 feet in fading visibility, dicey. Crews were seldom bored on these attacks, even as high as 500 feet with radar they were rarely taken for granted. All it needs is a piece of equipment to fail or give a wrong reading: disregarding a cue as false is no sweat at 1000 feet but lower down in the dark it's another matter. A simple failure can be fatal even to an experienced crew. Barry and I had too little Buccaneer experience to be complacent. Wary of dangerous disorientation in fading light and high speed over slack glassy water we were developing that suspicious nature that is the hallmark of a sound Bucc crew. Down to excellent instructors, I might add.

Both of us enjoyed pushing boundaries but it was never reckless – more controlled belligerence front-loaded with bags of humour. We were at that point as a crew where we felt we could take anything the OCU could throw at us. Our combined experience from completely different directions – mighty V- Force and single-seat fighter had somehow developed into a balanced consistency thanks to the OCU. They wanted it to be a tough course and rightly so considering the environment we flew in and we cheerfully hoisted on board all their comments. Well most of them!

Part of any success we enjoyed was down to our wholehearted absorption of the 'Buccaneer spirit' and robust crew attitude in flying the aircraft. We never took ourselves seriously for a start. We weren't aces. But then again we hadn't joined the Buccaneer force to be bored so we considered all OCU staff as fair game to bounce and beat. One instructor dryly commented, ''You two threw yourselves into throwing back everything we bloody threw at you'...

Soaking up advice from the instructors we unmercifully badgered any with worse scores than us. Especially if we detected a whiff of narcissism – rare on the OCU. A hard honest de-brief was essential: there were no one v one air combat sorties, it was all ground or maritime attack but I loved a scrap and whenever strike progressions were flown Barry and I noticed a humorous wariness creeping into debriefs. The staff were sharp, cottoning on that silence on our part wasn't a weakness – it was the start of our revenge – we were always amicable in the bar after incendries.

On the 10th of November, I flew with Lieutenant Ritchie on a night laydown attack on a ship target at Wainfleet Range, he pointed out various dangers and everything went fine. But on the 11th, the next day a crew from 809 attacked the same target as dusk settled on the range and crashed on the mudflats. They were an experienced crew recently disembarked from *Ark Royal* who would have been worked up and fully operational. The failure of the radio altimeter was to blame – there was no modern standby instrument package on Buccaneers, unlike today's jets.

The ejector seats fired on impact when the aircraft broke up. The pilot Steve Kershaw tragically died and the observer Dave Thompson survived entanglement in his parachute underwater and would have perished in the cold but for the gallant efforts of two men manning a Boston dredger and a pilot boat who managed to get him clear.

That crash made me sit up and take notice for the rest of the course, It happened to a good crew, a steady experienced pilot with an equally steady experienced observer. Accidents don't stop you from being punchy but to be honest they do help stop any slide towards complacency if you happen to be doing comfortably well.

The pace quickened; we fired rockets and I flew pairs dive-bombing with Squadron Leader Pete Eustace which was fun. Eustace was an experienced hand and easy company, he had been a navigator on Shackletons which straightaway drew my respect, he must have been outstanding to take the road to a fast-jet navigator and instruct as well he did.

So far we hadn't cancelled a flight through aircraft unserviceability until I foolishly boasted about it over a beer. Murphy's Law promptly struck at dusk on the 19th returning to land before another sortie on night long-toss – I couldn't get the nose wheel down. Barry took me through the emergency drills and we eventually got it down, lowered the hook and took the RHAG – Rotary Hydraulic Arrestor Gear – as I had lost hydraulics and brakes. It was a no-sweat gentle arrest but the downside was sitting with the fire trucks to unhook from the wire and listening to choice comments from guys in the circuit mad about me closing the runway making them late for the bar.

Afterwards, I met Danny Daniels, quickly briefed and belted off to the range in another aircraft for my first look at night long-toss bombing. For a first attempt in the inky pooh at 500 knots pulling up into the attack and then over-banking to pull back below the horizon (if you had one) is quite bunchy stuff. So to be honest on return I felt a bit knackered when Murphy struck again at my hydraulics.

This time I couldn't get any landing gear down at all but after a bit of fruity language on my part and Danny running expertly through the emergency check-lists, we sat hooked in the RHAG again surrounded by fire trucks with blue flashing lights. Sarcasm from guys circling while I blocked the runway making people late for the bar twice in one night fell on deaf ears, I had no hope of getting there before it closed either.

The next night Barry and I ignored smart remarks about my record of hydraulic failures and went off to fly night medium toss which we discovered was much bunchier and sweatier than long-toss bombing. But life had a feel-good factor with the end of the course in sight and I naively remarked to Barry nothing could go wrong now; I had experienced more hydraulic failures in one night than most had in years. I was almost right – we made it to the end of the month before I blew them again!

The last trip in November was a pairs attack in France with an OCU back-seater and wingman. But fate had it in for me. We block the runway. Again. It was only a 'no nose-wheel down' this time and I was getting good at hydraulic problems so we whizzed through the drills and took the RHAG. But I was a bit miffed and forgot to take my foot off the brakes on arrest and blew a tyre with an astonishingly bang! Peeved to hell at my clumsiness I sat in the wire amongst blue-flashing lights and milling firemen waiting for a tow-bar but with skin now rhinoceros hide thick, I just turned the radio volume down to drown out the sarcastic comments of the course arriving back short of fuel.

The feel-good factor returned directly after we met our next boss Wing Commander Pete Rogers OC 208 Squadron. Barry and I joined him surrounded two-deep at the bar with a noisy bunch of his crews. He introduced us to our new flight commanders Pete Jones and Graham Pitchfork who gave us a warm welcome and mentioned that Barry and I could stay together as a crew if we wanted. Great news. Barry was not only a first-rate navigator who had done well on a course for which selection had been tough but importantly, he was the tact and diplomacy part of our crew. I was lost without him. He agreed! Brilliant!

True to his word the CI Graham Smart asked me how I felt about the Buccaneer after finishing. I told him he was spot on. In a short time, I had developed a great respect for what it could do and I felt the staff had been suitably punchy and professional in line with the aircraft. He grinned and shook my hand and told me that Barry Southwould and I were rated as 'Above Average'. Boy! that was unexpected. Probably wasn't the right moment to ask if we were getting AIM 9L Sidewinders to replace the Navy's crummy old AIM 9B's either. One thing for sure, any success I may have enjoyed was due entirely to Barry Southwould's strong performance as a fast jet navigator.

Jackaroo cockpit

Thruxton Jackaroo, photo Alan Wilkinson

30 Squadron Sharja Trucial States

30 Squadron Empty Quarter 1960s

Nigel Voute Bill Burborough Chris King Phil Flint Author John Bartholomew Steve Riley Steve Jarmain Ian Davidson 1970

Wings presentation

Kenya Job done Left Ivor Gibbs CFI centre

1970 Ian Davidson on the ascent

Advanced Flying Neil Matherson Author Steve Riley Gus Crockat Nigel Tingle Phil Flint Dereck Poate

Daddy Probin's home build minus canopy lost over Mt Kenya

Image © Michael Rondot collectair.co.uk Pilots loved the Hunter

58 Squadron Hunter

45 Squadron West Raynham, Squadron Leader Wally Willman third from left Callum Kerr on his left

Painting of 208 Squadron Buccaneer by Michael Rondot captures aggressive fighting qualities and rock steadiness of the aircraft

Winners of weapons trophy Author and Barry Southwould

Gilroy Weapons Trophy held by Phil Ford Graham Pitchfork and Mike Bush with early shakers and movers commanded by Pete Rogers who maintained the squadron's formidable reputation

Image by James Biggadike, A welcome basket streaming for you

In the Weeds

HMS Ark Royal ' Mother' Ark collection

Loading onto a catapult is a disciplined affair

Buccaneers are the cutting edge of the carrier, Authors Ark collection

On sight, Author

Race Day high standards of dress and behaviour FOCRAB Bob Joy, Author and Tony Ogilvy right the racing commentator discuss suspected knobbling by Author centre

Gun Deck Goofers, Photo Pete John
Paul Barnard USN – 809 LSO & Anwyl Hughes spotting firing on splash target

photo Pete John, Lt Cdr Tony Ogilvy Weapons Leader front with 809 boys.Anwyl Hughes South Wales Borderers centre right

Ops Ready, Dick Aitken, Ed Wyer, Author and Bobby Anderson

Brian Hoskins leads 1979 Team – brochure shot Richard Cooke last year of Gnats

Dynamics of close line astern 1979 Brochure

Authur Gibson's First photo of new Hawks with Concorde

Wineglass formation Brochure First Springhawk

Prince of Wales Feathers Arthur Gibson

Boss with HM King Hussein in our new Hawk

Byron and Boss breaking bad news to Crown Prince he may have to fly with the Author

The following contemporary winter training photos by Claire Hartley her blend of smoke light and shadow adds an extra dimension to the Reds imaging.

'Tightening'

'Rolling'

'Big Five'

A most famous painting 'The Red Arrows' © Michael Rondot

'Red Seven' © Michael Rondot

'One more to go'

Team 1982 Tim Miller John Myers Author Ian Huzzard Lead John Blackwell Byron Walters Tim Watts Henry de Coursier Phil Tolman

Ian Botham cricket star who has faced the world's fastest bowlers and a keen flyer himself photographed with Team Leader John Blackwell after flying with us in 1982 season.

It's been a privilege

chapter Twenty-Nine

208 Squadron

EVERY PILOT wants to be part of a squadron with a fine history and I was extraordinarily lucky to be appointed to 208 Squadron with its distinguished past. What's more, the two original commanding officers were alive when I joined it. A browse through the squadron history reads like something out of 'Ripping Yarns' of yester-year crammed full of brave deeds and accomplishments

The original squadron boss Geoffry Bromet handed command over to Squadron Commander Chris Draper who not only led the squadron well shooting down many enemy aircraft but made something of an industry in night bombing attacks on enemy airship sheds. On one occasion Draper burned all the squadron's aircraft when fog prevented flying out in the face of an enemy advance.

The Press christened him the 'The Mad Major' in 1954 after flying through London Bridge in his seventies. The exploits of such men I had read about as a schoolboy. Yes, 208 was the squadron for me but I obliterated all thoughts of flying under bridges.

Barry and I, Phil Ford, Jim Crowley, and Trevor Brown joined together and discovered we had a superb Boss in Pete Rogers who led from the front. His executives were all good with a laid back confident air. The responsibilities of flight commander rested lightly on the shoulders of Pete Jones and Graham Pitchfork and for our Squadron QFI, we had a real ace in Bill Aspinall, late of the Red Arrows. Our Weapons Instructor Mike Bush made sure we not only delivered weapons accurately but enjoyed every minute of it.

Our role was to support land forces resisting an advance by Warsaw Pact forces into West Europe with conventional munitions and then with nuclear weapons in the event of escalation. An overland role was fine by me but not one flying over grey foggy German plains or down a rain-drizzled Ruhr. No, that was for the likes of 15 and 16 Squadron, massively into missing high masts in lousy weather and allegedly sharing heroic bullshit-sessions in the bar outdoing each other – But held in great respect by other air forces.

The Americans liked the better bit of Germany with skiing in Bavaria and Munchen Beer Fests served by lederhosen-clad Bavarians.

But that wasn't for 208. We got to fly over Norway and North-western bits where you could see where you were going. A place packed with glaciers and fiords and bloody good low flying where you missed elk and reindeer by a whisker and the fishing was terrific. Where friendly Norwegian waiters and gorgeous waitresses with really blue eyes helped you to big white plates full of pink prawns and you drank deeply of their expensive beer. No comparison. We won hands down for a job description.

But reading between the lines one sensed a darker side. When the Soviets struck they would be all over the Scandanavians and their airfields in no time. This meant we would have to bomb the hell out of all our friend's airfields. If this failed, we go nuclear.

Of course, this was all secret. And although aircrews can be surprisingly tight-lipped the RAF had to be sure. They didn't want a last-minute show-stopper so we were Positively Vetted for high-security clearances, because, coming our way were large folders with 'Secret' and 'Top-Secret' plastered all over them.

It was routine to be negatively vetted – a police check that we had no record of criminal charges for un-officer offences such as arson or burglary. But positive vetting was a new ball game. This was carried out by ex or current counter-espionage chaps rumoured to be stone-faced and didn't miss a thing.

I wondered how far back in the past they would dig. My mind was already conjuring up ghastly accusations they could lay at my feet.

Maybe a young lad with a penchant for excessive use of black powder bombs and unnecessary large calibre guns didn't point to being responsible. Burning down a sizable Welsh hill of dry gorse and fleeing ewes probably wasn't the best testimony for someone the RAF intended to hand over a nuclear bomb to either.

Any person flying a role contested by academic circles might be apprehensive about explaining a nuclear doctrine. But I rarely looked into the murky waters of left-leaners. No! I was perfectly happy flying the Queen's fighters defending everyone.

Compared to a non-flyer my politics like most aircrew were uncomplicated with a cultivated whiff of political immaturity to preserve a simplistic view. We had a perfectly good House of Commons that functioned adequately for centuries stuffed with politicians who would let me know who I was going to shoot at. Why worry? Besides I had other concerns. I was hoping they weren't going to ask my father if I was a secret Marxist. That would be a career-buster. Being a consummate Royalist faced with a pushy interviewer he would just reach for one of the shotguns on our beam – the one he always kept loaded with a heavy cartridge for fox...

chapter Thirty

'Flying Shuftis' Badged Operational

TWENTY-FOUR NUCLEAR weapons were allocated to 208 and to be on the safe side we had to be badged combat-ready in case the Prime Minister asked us to use any. The squadron crest we badged with was adopted in 1930: A symbol of a winged eye looking out of an azure blue sky signified the squadron's recce work in the Middle East.

We got cracking, I flew a check ride with Graham Pitchfork finding him a kindred spirit as soon as I let rip with some aerobatics and general throw about. He was enormously experienced on the aircraft having flown with the Navy as well as RAF and with no two-seater Buccs there has to be a check ride. Fine – as long as there was no undue criticism of my flying. I might be a new boy but single-seat blood still flowed in my veins but I felt comfortable taking his advice and felt we were a crew the minute we strapped in. Bursting through a thin layer of cloud into brilliant sunshine for a throw-around I couldn't think of a better way to start a trip to Scotland: and all I had to do was behave. And I did. Over a beer, Graham made it clear I was welcome; no waiting, no reports to chew over – the sign of a good outfit. His enthusiasm hinted strongly at similar canons to me on flying issues which meant good flying ahead (I was correct in my judgement because Graham eventually commanded 208 Squadron)

In January '75 Barry and I began a full-on work-up to combat-ready. 208 was dual-capable – Strike and Conventional.

Strike missions were flown individually to hit targets deep within enemy territory. Attack missions would be flown in formations to provide mutual support from enemy fighters. We started the Strike bit first. This entailed fast first run attacks (FRA's) on a range with a time on target (TOT) where Barry worked out a take-off time. And it didn't take him long to shine.

Working together meant he did all of it because timing techniques were known only to God and back-seaters and jealously guarded. Navigator skills in getting these attacks down to the second were so good that pilots tended to take it for granted and I have to admit that compared to some of the shambles I led on Hunters, they were brilliant. The precise brakes off time I worked out often bore no resemblance to reality. The calculations were accurate enough but I seldom left spare time for contingencies which didn't sit well with my inclination to pick a fight with another formation. (Cyd Sowler used to beat me over the head with a rolled-up map for this)

Considering the volatile nature of some of the Warsaw Pact countries the RAF rightly took the prevention of nuclear war seriously. They had good reason to. Starting Armageddon was a lot easier than preventing it. However, starting World War Three in the RAF was not a spontaneous affair! No. We had Strike Planning to do. And I wasn't alarmed by any lack of moral rectitude – we all knew 'they' would start it all so I wasn't fussed at all about blowing up a sizable part of the planet with 'them' on it.

The RAF handled anything nuclear with commendable thoroughness. Allocated targets lay in distinctive red Strike Mission Folders with TOP SECRET stamped all over them. Not only were our targets Top Secret but our profile and routes were also secret including known missile and heavy gun defences standing in our way.

All Strike Folders were kept in a secure room called the Vault. I expected to be confronted with combination codes and lots of twirling levers on a big shiny steel circular door like the ones

Hollywood villains spend days trying to drill through or dynamite open.

I was disappointed to find it wasn't like that. An ordinary boring looking steel door confronted us in Operations. Even the peep-hole failed to impress.

Access was restricted to a few officers with high-security clearances and everything was strictly 'Need to Know'. The Vault Officer was armed with a 9 mm Browning to ensure we didn't do a runner with our stuff and after checking our IDs he opened our Strike Folder and slid it across a huge map table.

The first thing that struck me after a sharp intake of breath was seeing exactly what I was going to obliterate.

The second thing was the sheer scale of Russians, Warsaw Pact forces and heathens that aimed to obliterate us before we did it to them!

A map showed our planned track through a maze of different coloured dots. These dots signified Surface to Air Missile (SAM) sites, gun sites and fighter defence airfields plus a few hundred anti-aircraft radars. And it got worse.

The chilling words across our inbound route ''Switch off radios'' meant no recall. It was one way. We were now unstoppable.

I don't mean a suicide attack because I knew for sure the RAF wasn't into white headbands and pre-kamikaze saki bouts before take-off. This was more subtle. I was un-phased by 'Jettison Tanks' marked on the chart which was standard knickering off empty external fuel tanks to reduce drag. But! After surviving a maelstrom of defensive fire (probably for several minutes if we were lucky) two large words caught my attention. They were slap-bang on the return arc over the permafrost of Finland.

" EJECT NOW".

I toyed with asking if there was any chance of hooking up to a tanker on the way home. But a warning flashed across my mind about 'Staff Officer Solutions'. If a Tanker was available they would give us a slot all right – but on the way in. That way we hit further East, they looked brilliant planners and we run out of fuel even further away.

Barry being ex-V Force casually shrugged it off and I tried to give the impression I knew all along something like this was expected. There wasn't a shadow of a doubt we weren't up to it of course. The inspired optimism I took from it all was that we were bound to eat a lot of reindeer on the thousand-mile walk home.

Practising our war profile in a simulator is noisy with electronic countermeasures (ECM) shrieking warnings of missile launches and fighter interception radars tickling us. Our final run into target was memorised as planners rightly felt we would be busy and because you could guarantee the Russians would complicate things by starting a war in the dark: night laydown attacks were next.

Bill Aspinall checked out my night instrument flying in our Hunter.

When I say checked: I meant 'frighten the living s**t out of me'. On a night as dark as Satan's armpit doing 420 knots at 200 feet over the sea making steep turns on instruments is challenging enough and with no radar, even more so. But then you always trust the guy next to you because he can see out. When you know he can't see anything either it puts a new angle on everything.

Bill, being ex-Red Arrows knew a thing about accuracy and was brilliant on instruments; upside-down or tight on someone's wing he could still read the dials so wimping out was not an option.

As for the simulator rides: we stopped twice when black diplomatic cars from the Russian Embassy plastered with listening gear, aerials and dark windows parked outside the sim' on the public road…

The Russians were taking us seriously: We badged combat-ready.

chapter Thirty-One

Tanking up

OFF-DUTY LIFE in the mess was lively with expensive mess bills. Given that aircrews are the necessary stars that animate the place, burning pianos and damage to furniture from wild scrums after a dining-in night tends to add up and with two active attack squadrons trying to outdo each other life wasn't boring. Being new on 208 didn't last long with more guys arriving so I gave up on any tendency to keep quiet and gave in to the heavy bar gravity drawing us into happy hour. We were a happy bunch and having a QWI like Mike Bush, a Boss like Pete Rogers plus two punchy flight commanders in Pete and Graham shaped the squadron's aggressive outlook. With the North Sea getting crowded with oil rigs: it was inevitable for rig protection patrols to start. Piece of cake for us because our practice bombs were perfectly capable of sinking any nutcase trying it on in a fast zodiac rubber boat. For the serious terrorist, a pod of 2inch rockets would take care of any larger fast vessel much more stylishly.

Boss gave me a lead position early which was brilliant. One morning I ventured to Mike Bush we should try something different faced with bad weather in deep valleys. Splitting into two pairs wasn't a problem but why not stick together?

The result was everyone laughing at my expense when I tried flying wadi-arrow like the Aden and Gulf boys did in steep wadis. An arrow formation of 75 yards between Hunters might be OK but was a shambles in the faster Buccaneer. Guys kept hitting jet-wake and complaining so I shoved everyone out to 150 yards and then wound it up. But I had to abandon it because of cries of anguish and derision from behind. Calling them big girl's blouses didn't help my case much either at a rowdy debrief.

So from the ashes of a good idea the only thing achieved was a nickname for me – 'Wadi Ward'. And it stuck!

Weapon training out of Deciomomannu airfield in Sardinia at Frasca range was a buzz, the weaponry almost as intense as the wild drinking in the Pig and Tapeworm bar in our block. The Germans paid more for Frasca Range use so they got the lion's share of the slots, the Italians being as tight-assed as the Brits got less and their Fiat G91s were frequently idle. The Germans used F104G Starfighters and we Brits cycled all our ground attack fighters through there; but unfortunately, the Italians controlled it. They were nice chaps but the range helicopter had appalling serviceability and we frequently heard *'thee helicopter – she no-fly – thee range she is closed'!*

Germans Starfighters weren't good in the ground attack role and they lost a few which was unfortunate. One crashed, closing the range for us and killed the pilot; the board of inquiry were mounting the steps of their transport to go home when another Starfighter flew overhead and promptly crashed: so they turned around and reconvened … Despite everything, Barry and I did better than we thought ending very near the top of the weapons ladder.

Norway became part of our parish with frequent overnights and days walking in the mountains. Doing anything in Norway, drinking or flying was always fun and flying low over snowy terrain in Norway I found a new way of frightening myself.

High icy plateaus stunted with trees in blowing snow is a problem. The trees all looked the same whether they were normal-sized spruce or small Christmas trees size so it's difficult to tell their height. And this is where the habit of just missing the tops can get you into trouble.

Leading Bill Aspinal and Jim Crowley on a pairs sortie over a high Norwegian plateau I nipped over a tree that I thought was quite tall and was amazed when our rad alt red pinged off and went to twenty feet just before I flew over it! With frequent white-out conditions, it was sporting stuff especially when our Nordic friends put up F5s to intercept us. Another danger was powerlines strung over fiords. Fiords are particularly enticing to all flyers but it's difficult to see these wires easily and sadly a Buccaneer from another squadron hit one. The observer survived but the pilot was killed.

Air to air refuelling next! Something I was looking forward to.

Ken Mackenzie a respected Naval Looker doing time with the Crabs was going to fly with me for my first prod and to get in the swing of future night tanking I flew a night formation check with Bill Aspinall. It made sense because we all flew night weapon sorties at this stage but we rarely flew in battle formation in darkness at low-level – 1000 feet was the norm overland at night to avoid hitting wires. Bill decided low-level 200 feet at night in tactical formation was good character-forming stuff for me and vaguely promised it would be over the sea – getting caught in competitive scenarios comes far too easily and I stupidly remarked to Bill he could wind it up a bit. Mistake.

Bill didn't hang around. I raced to line up alongside him on a wet runway thinking the weather wasn't brilliant; reflections from the runway lights cast distorted arcs across inky dark puddles and the lights stretched ahead in a series of white blobs blurring into the rain and mist. Yes, bloody mist!

No nav-lights or anything Jessy for our Bill. No, this was big boys' stuff and I was the one with an anti-collision light on and its stabbing red flash added a demonic dimension to an already dank scene. All I had to look at now were tiny purple rearward-facing tactical formation lights. Take-off was smooth and rotation neat and I had no difficulty at all in keeping tight to the bulky dark shape next to me – until Bill's casual call: 'anti-collision light off.' God, it was dark.

Flying with the Reds had given Bill real insight into formation leading and no matter what speed he flew at he initiated a turn with a smooth take-up. But it was still beads of sweat stuff because in winter we wore immersion suits; with rubber collars, built-in rubber socks and inch-thick underwear. Especially sweaty if you are in tactical formation low down doing God only knows what speed in the inky-pooh. I couldn't see a thing except for a purple glow as dim as a nursery light somewhere to my left and I didn't dare look inside in case I lost sight of it. But I was learning fast – with one purple light I hadn't a clue of our bank angle but dropping back slightly I could just make out two; instant orientation and a break in the deep breathing you do when bunched up inside.

Breaking out in crap weather was not an option. Bill would just tell us to re-join – and you would have to be completely out of your mind to attempt that!

Without a doubt, it was the most demanding flying I had done so far and on landing my immersion suit smelled as high as a Sumo-wrestlers jock-strap.

Ken Mackenzie had sat behind many pilots on their first aerial refuel and their attempts hadn't mentally scarred him because he showed no sign of doing a runner when told he was flying with me on 208. He gave me a thorough brief of what the basket would do and I liked his easy-going manner. I found later this was the best approach to this particular skill; if you made out it could only be accomplished by a ten-foot-tall ace Naval pilot with a big watch it was possible to feel outraged. You then get bunched up and clumsy and I was agricultural enough already.

The plan was straightforward. We would use an OCU Bucc tanker borrowed from 809 NAS and Lieutenant Ken Mackenzie RN, doing time with the Crabs on 12 Squadron sat in a 208 Buccaneer behind me providing excellent advice somewhere over the North Sea at 10,000 feet.

To the Navy air to air refuelling was basic stuff and if you couldn't do it you simply fell out of the sky if a lift stuck a few feet below deck level made hooking on impossible and called for a tanker launch while it was fixed. The refuelling pod could be fitted to any Buccaneer like the one ahead of me streaming his hose and basket. Buddy-buddy refuelling between Buccaneers is just accurate formation flying but procedures have to be followed as a few things can go wrong. Refuelling from a Victor or larger aircraft can be dicey at night or in turbulence so discipline was vital.

Ken filled me in on a few Fleet practices which convinced me all FAA pilots were as dishonest as me when it came to declaring fuel states. A formation returning to 'Mother' down on fuel normally follows the 'lowest gets it first' rule but this can lead to telling porkies if you have a diversion ashore with a dry bar.

And if you had a Scots back-seater giving fuel away you can get short-changed in the blink of an eye.

With a big tanker like a Victor it's normal to join on the tanker' port side; that way the captain can see who he's giving expensive fuel. Generally, two hoses are in use but sometimes three and when cleared you simply move down and across to position behind the basket streaming from the allocated hose. Once refuelled and cleared by the Victor tanker you carefully move into echelon on his starboard wing to wait. Procedures were much the same on Buccs except there's only one hose.

Behind the hose and basket is where the real formation skill comes in. Every type of new jet will go through a test pilot's hands to establish an airflow effect on probe insertion. A position is then established on the tanker's pod to visually place the basket as you move up the hose line. Ken told me to visually place the basket in the pod's right bottom quadrant – 'AND KEEP IT THERE'! It meant we eased up the hose line to a point where the slipstream cleanly moved the basket away into line with the probe. It's very close and the impression you get is the basket jumping up and to the right at the last moment.

Ken run through common mistakes for me. The basket will move around – nothing is perfectly still in the air. In daylight, if you focus on it instead of doing what you were told and ignoring it you end up chasing the thing. Flying through turbulence doing this won't help either. A miss guaranteed!

Hitting the rim can be disconcerting but providing you are slow it's OK. But if too fast and aggressive this feeling quickly changes to humiliation when you break your probe off. Nasty!

'Spoking' is where you have a bit of yaw on for some reason best known to yourself and your probe bursts clean through the flexible metal struts on the sides of the basket which gives it rigidity. This can be frustrating but if you're careful backing out it's no big deal. However, if you screw up and come out fast with yaw on it's another matter.

Hoses cost a lot of money and the Cobham pod hose will come out of its housing all right: but it will not plummet into the sea on a little 'chute as you might expect.

No. They come out and wrap themselves around your aircraft like some demented Triffid.

(I know this because I saw our naval friend from training taxi in with one wrapped around his aircraft and draped over a wing)

'Cleared contact' came our crisp clearance.

'Roger cleared contact' I replied and easing power on flew in with the walking pace of an asthmatic ant.

'Slightly faster Wadi.'

Further back the basket looks relatively small and easy to fix in position but closer in the relationship alters – the basket gets closer and larger so you have to compensate by focusing on the pod ignoring basket movements.

Ken was a consummate professional with little chat but continuous with it.

'Nice – nice – nice- good speed keep that Wadi – Contact.'

'Fuel flowing.'

'Fuel flows.'

After three more good prods, we broke away and merrily tail-chased home feeling chuffed to bits. It was all luck of course but I took away valuable advice from Ken. Stay in position. Keep calm. Stay loose on the controls and never do anything in a hurry. Brilliant. I was to remember his advice when I ran into a spot of trouble behind a Victor tanker.

chapter Thirty-Two

Gearing up

TWO WEEKS after my first prod in October 75' I flew behind a Victor Tanker for the first time as two of a pair. Behind me sat Ken MacKenzie prised out again to give me advice. We met the tanker on its towline – a racetrack course over the North Sea where she flew ahead of us looking huge against the crystal blue morning sky.

Overtake in the final stages of a pair join is about 40 knots which gives time to orientate yourself. This might seem slow but we were feeling our way into a deeper role; one with a faster pace. Tanker crews were highly professional with procedures to refuel a formation at night sometimes in radio silence which is risky enough. But the eventual aim would be to climb from low-level for a night tactical join then refuel in the dark and kiss-off descending high speed in tactical formation to attack at first light.

208 was fully into the heavy attack game now.

With a four-ship, wingmen have to stabilise on the join which meant a good power setting and careful use of airbrake so they didn't slide past: there's a lot of momentum with a heavy weapon load and it's easy to screw up if you forget.

Buddy-buddy tanking lowdown is a piece of cake compared to tanking at higher levels.

Up high you never lead someone into the turbulent flow behind the tanker's wing. Even a simple turn at the end of the towline can bring problems: three men on the wing called for steady handling because the turn is towards you. It's a doddle in daylight but not in the dark because it's difficult to judge drift towards you when the tanker's wing drops for the turn. Thanks to Ken I checked out daylight tanking as a prelude to night operations where we would do it all in radio silence. No point in making life easy!

Just how risky air-to-air refuelling can be showed up with a tragic accident earlier in the year. It came out of nowhere during daylight training. One of the best pilots we had with probably the best back-seater in the force got into trouble. The tanker received a hit on the tail and it was all over in seconds: the aircraft disintegrated and only one of the five crew survived. 57 Squadron Tanker boys were professionals regarding the accident and their crews showed real resolve getting up again despite their loss. The same grit was shown by the Buccaneer crew: with the full dangers of their role so recently revealed they not only had to go up in the day to qualify but then had to do it at night. They earned the respect of the whole squadron in the way they mentally and physically handled it.

With less performance, refuelling at 30,000 to 35,000 feet brings problems. Flying slow behind the tanker doesn't help when taking on a large weight of fuel on top of our weapon load because the marked increase in weight brought us closer to the stall which called for delicate handling. I never got bored tanking!

Taking on fuel gives extra time aloft putting you close to your bladder limit. And nothing worse than wetting yourself in heavy underwear. The RAF's answer to this rather personal issue was a flat plastic bag with a long neck containing a compressed dry sponge. This might seem useful in time of need but connecting yourself was daunting especially in turbulence close to the ground.

I did it once and got heckled for being out of position after almost colliding with a large granite outcrop in Scotland just to have a pee. After that little episode, I decided it was safer to drink less coffee before take-off.

208 came up to strength in 1975. Flying more complex sorties, pilots made quick judgement calls; I experienced this in Scotland with Roger North on the wing and Ricky Ramjet number three. After a simulated attack with a hard pull-off, the weather closed in rapidly – and I didn't have to say a thing. Within seconds they were rock solid on the wing; I found a gap, dived through it and they were out in a flash giving cover.

Routine stuff maybe but without a word spoken, it reflects well on everyone. Both thought it perfectly normal.

I asked the Boss if Ken Norman and John Broadbent could be deputy lead to us; they were a great crew that could think on their feet with a similar outlook to us (refined belligerence with a hint of style!) First time out as a four-ship we got into a running scrap in Scotland with a two Phantom bounce. Staying high-speed down in the weeds below cloud we frustrated them with hard turns where Ken was a natural, piggy-backing the lead with me keeping them in and out of burners to get some overtake.

The pace quickened and I flew my first night-tanking sortie leading a pair onto a Victor tanker. To assist our prods the basket has beta-lights around its rim like dim highway cats-eyes which are not obvious in daylight and at night you have to be close to see them.

With the tanker lit up like a Christmas tree night refuelling is easy but it's a different world without his lights. Lacking an interception radar, finding him wasn't that easy and stalking a dark shape in night tactical formation made the hairs on your neck stand on end. Barry always kept his cool but I was highly aware of unvoiced concerns from the guys on join-ups. It's that paradox again of relaxing with fingertip control but being vigilant to the slightest degree of movement towards you. During early tactical night tanking I was a ball of sweat. On one night plug, an engine surged – probably my gross engine handling again – which ordinarily would have been a mere nuisance but you don't muck about doing drills. No Sir. You slowly and gingerly get the hell out of the basket and slowly and carefully bug out of formation! Later the engine subsequently faltered and we shut it down.

Developing tactics against fighters with our new Westinghouse ECM pods – electronic countermeasure – was entertaining. Our pods were brilliant. Designed to trick or deceive radar or other devices: they could be offensive or defensive with intelligent jamming or barrage jam by making senseless noise to blot everything out. In protecting us from guided missiles they could present many fake targets, make us disappear, or even move about at random. Graham Pitchfork and I flew sorties building up basic tactical manoeuvres. First was a Lightning using his radar on a pair of us at low-level. Any decent ECM will pick up a fighter intercept radar before he picks you up (your received signal is better than his returned one) Immediate wide battle formation makes him commit and the 'zip…zip' of his radar shows clearly to get a bearing, but you have no idea of his range. Graham called headings and we put him at ninety degrees and accelerated.

The attacking fighter now has to work at it because his radar is not so good on a target beggaring off sideways. Particularly one doing 540 knots plus and he is using up his fuel fast. These tactics are used with ordinary ECM fitted to your standard drive off the garage forecourt Bucc, but, with our new pod, you can steal his radar range gate and retransmit it to put you miles away. The essentials I took from Graham were the navigator runs the fight or evasion from the back seat and full cooperation is needed by the pilot to make it work. In a wide battle formation of up to ten miles often zipped lip radio silence it required a good grasp of the third dimension. Hence the expression 'fightigator'. A name that sums up the role of a fast-jet navigator rather neatly.

The golden rule is you are quids in if you avoid visual detection.

It got better against Phantoms. They have pulse doppler radar that can pick you up at long range but are noticeably poorer with a crossing target at right angles. Blast off goodies with the pod warms your heart to see them in the far distance taking pot-shots at holes in the sky to boost their Brylcream image. With Graham choreographing back-seaters on ECM usage our battle formations remained undetected more often; such processes can't be hurried and it's fun with a steel edge.

Increased range from tanking put us in a better position for coordinated attacks from different directions. Squadron execs were always up for new ideas and I'm convinced that this attitude framed by Boss Roger's confident command put us into a more dangerous category of a foe.

The humour and tempo of two-man crew briefings made me chuckle: it's where the robust nature of Buccaneer ethics are embodied with a humour that's a hallmark of the force. Briefing for an eight-ship coordinated attack involves sixteen guys dressed scruffily in immersion suits. Underneath they have green-pile underwear an inch thick to survive an ejection into a cold marrow freezing North Sea – which means keeping zips wide open to keep cool on the ground. Their ejection garters make clunking noises as the guys bounce off corridor walls and the banter is so boisterous they can be heard well before being seen.

Clutching steaming coffee they file noisily into a room that always seems too small.

In a corner stands a whiteboard covered in scrawly diagrams where your web of arrows pointing in all directions is alongside nice neat lists of timings made by the lead navigator. And nobody listens to you at first. Ribald banter and hurled insults of cockups the last time up, take a minute to subside...

Briefs are punchy and short. Charts are already prepared and targets studied. Every lead will stamp his style and footprint on the sortie whether he likes it or not. Usually in a casual but firm manner where notes are made, watches set and on cue guys shamble out in twos. Signing up they stroll towards a line of menacing shapes to check the outside then climb ladders set into the side of gleaming fuselages and step into cold cockpits. Minutes tick silently by; the rumbling growl of powerful engines starting in unison fills the air and the smell of burning kerosene drifts in the wind. The lead aircraft eases forward, his nose dips as brakes are checked, wings hung with an array of weaponry unfold with a flourish. The attack formation follows one by one, the muted rumble of their engines charges the air with a hint of restrained power as they move in a long camouflaged stream.

Fast forward a few hours and the same guys clunk in with smelly immersion suits zipped open, some semi-disrobed with arms tied around waists. Clutching coffee they look far scruffier now: dishevelled hair and oxygen mask lines etched on faces hint at the sweat lost.

Only this time the banter has shifted. The buzz is a loud questioning of parenthood and the inability of the other formation to see the f*****g target your ninety-year-old grandmother could hit with her eyes closed. Derision from the others meets the remark in a rising cadence. But it's easier to grab attention now – 'OK guys!' – a showy glance at watch – 'bars opens so let's get this debrief on the road eh.' Instantly all chatter ceases...

Such is the brotherhood of a good squadron. Between these briefing performances, the formations have skimmed the waves into the high latitudes of the North Sea. Splitting silently into two they head for their start points and the Westinghouse ECM pods give a false position to defending Phantoms. Back-seaters work at timings; front seaters work at flying correct speeds wary of hitting the ground as they give sharp cover. This is where the speed becomes apparent as your number three blurs against the backdrop of trees, moor and mountains. Twenty minutes pass and without a word he arcs away with his wingman.

Final acceleration. Throttles ease up. Ident point flashes underneath. Start now! Stopwatch thumbed. A rumble as you flick open the weapons bay; map's in your hand – the attack track you have memorised – a shape abruptly looms onto your wing – pull up now. He pulls up with you to simulate tossing eight 1000 lb HE bombs to airburst over the defences. Rollover, pull hard, miss the clouds and rolling back pull smartly to miss the ground.

Out of the corner of your eye, you see your number three: he's low down and fast with a shape glued to his wing. They tip in for a pairs dive attack timing to arrive after your bombs explode. A quick hard reversal to parallel them and they are in battle-formation with you doing 500 knots. Forty seconds after the hit the sinister darting shapes of the other attack formation are over the target; diving from a completely different direction.

But you are on your way home.

At times like this, it's easy to see why Bucc crews get bored and get fined for speeding in fast cars.

An upside to fighter affiliation was flying with the opposition in dual-seat Lightning and Starfighters and they with us. Good stuff. Flying F104s might be exhilarating but they were strictly straight line and you couldn't turn any harder than a brick with wings. That's what I told them anyway. The Lightning was lethal at any height but for close-quarter dog-fighting give me the old Hunter for a shed load of fun.

By now I had a healthy respect for bad weather and it showed when Al Ferguson and I led a night four-ship tanking sortie. Al an experienced hand was a slim-fit Liverpudlian who was inordinately proud of flying Buccaneers and walked with the punchy strut of a prizefighter. He agreed with me it was vile so I cancelled our tanker and brought the boys back. I suppose the confidence shown in me by our execs made me more assertive: buggered if I was going to put chum's lives at risk for the sake of my pride.

We all had enormous confidence in our aircraft; the robustness of the Spey was first-rate considering the seagulls and birds ingested and I only shut down two more – the one I wrecked on an engine air-test naturally doesn't count because there was something wrong with it anyway. One engine that blew was entertaining more for the casual chat than the actual event. A goose flying at 2000 feet over the sea was to blame and God only knows what it was doing there at night when we hit it. I felt a minor crump, a sheet of flame made a nice glow in the mirror and yaw to starboard grabbed my attention. Shutting down the engine was no sweat and I promptly got busy with various levers while Barry let me do my thing and sorted a checklist.

'Lead from two I think you have an engine fire – pause – You OK?'

'Thanks two but I'm busy…' 'Looks bad, you have a bloody long flame from your starboard side!'

'That's why I'm busy…'

'Roger that… OK, it's gone now …'

'Thanks, I'm going home. See you in the bar.'

'OK. Buggering off to the range that's two beers and two Glenfiddach.'

Re-scheduled to go up again the next day with Vic Blackwood we were cancelled and called in for a squadron brief. Someone higher up close to the almighty decided we should be flying lower.

This is a very silly thing to say to any Buccaneer crew.

We had to fly even lower

chapter Thirty-Three

Three Thousand Turkeys Never Made Christmas

EXACT POSITION skimming a pylon in Northumbria doing 540 plus knots

It was all legal. Farmers were warned of the low flying to come and the National Farmers Union waited for the claims to start. Being a country lad I had a slight miss-giving about the damage I might inflict on various stud farms, livestock and hen-houses; but this noble feeling lasted barely a minute when morning light softened fields and forests and the whole of Northumbria opened up clear in the sunshine.

Our aircraft were fitted with radio-altimeter recorders in the rear compartment that marked a 'trace' giving an average height flown. The boss was blunt at the briefing. ''If you feel sweaty overland at 150 feet – get back up to 250 feet''. We all chuckled – 250 feet was a high level nowadays.

We were soon convinced the height traces lied like cheap Taiwan watches. It seemed everyone felt the same and allowing for minor exaggeration our unofficial bar wash-ups summed it up nicely.

The radio-alt lights were hopeless overland. On the run-in to a target if the visibility was 'iffy' you had to aim for a pylon because you couldn't always see the wires. It's the highest point and all your concentration was forward avoiding the ground.

Someone pointed out that going *through* a farmer's yard was better than a gap between smaller buildings because there were no telephone wires (acceptable finesse)

But, our trace showed an average of something like 100 feet. Unbelievably high! We persevered, ridge hopping and hugging contours and it's amazing what passes through your mind once you get nonchalant about the whole business. No time for dreamy quixotic thoughts of lone oaks gentling the hillsides and forget poetic scents of wet pine and mulch on the forest floors skimming below. No! It was a case of hurtling a foot or so above a crown of trees hoping some huge mother of a buzzard wasn't going to lift off with a rabbit in its claws and splatter your windscreen – or even worse down an engine.

I idly wondered if you would write that up as a bird strike or rabbit strike: to my knowledge, no Buccaneer crew had ever reported a rabbit strike. But then again it could attract awkward questions about one's height.

I drew laughter when I claimed I saw someone fly *through* a forest (crew remains nameless to avoid future legal problems): he said the fire break looked too tempting and it was in the right direction anyway.

Because we were almost ahead of our sound I winced at cows with heads down placidly grazing and the bowed backs of tractors drivers oblivious to the clap of thunder about to hit them! A definite rule of thumb materialised: Flying at 30 feet meant too much concentration looking forward – poor lookout!

But the problem with toiling away at ultra-low level work is that it leads you innocently into trouble. Especially when taking a four-ship on a first-run laydown attack on a Wash range after a good thrash around with that warm fuzzy feeling inside that everything's legal.

On reflection, some of it *was* my fault.

I was probably too much in the groove and kept it a bit low. At 540K or so a turn is wide and four of us in a 30-second trail followed the same track. Scores were good and our recovery at Honington was stylish and immensely gratifying. Walking in listening to the sharp banter and witty repartee it seemed to me that enjoying yourself like this was almost a sin. But the feeling didn't last long.

Shambling heavy-footed in harness and sweaty gear for debriefing I was surprised when the boss stuck his head out of his office with his pipe going nicely.

'Wadi could you pop into the office please' he said casually.

'Sure Boss, is there a problem?'' I didn't get rollicked too often these days so I was curious. 'Not a big one but you've just written off a few turkeys.''

'Gosh! How many?'

Leaning back with his hands behind his head he puffed contentedly at the ceiling so I guessed it was bad news. He was always calm with bad news.

'It's the medium weights' he said, 'and for the record Wadi can you confirm it was you up north and on Wainfleet Range this morning.'

'Definitely, me plus three Boss.'

'Thanks Wadi the NFU and a turkey farmer have just been on the phone and we just need you to confirm for compensation.' He returned to the medium weights.

'The light-weights are OK you see, nothing on them and the heavy-weights are hefty buggers and slow. No, the medium weights have the wrong feather-to-weight ratio so when they panic into a corner they tend to suffocate themselves. And by the time the last aircraft went over the heavy-weights had joined in.' The boss was good at summing up.

'Err... how many?'

'Three thousand and counting.'

A twinge of conscience about so many dearly departed middle-weights prompted me to ask – 'would it help if I drove over to say sorry?'

'No! The farmer's quite happy the compensation will be hefty. Best leave it there Wadi'.

After such a reasonable tone with no hint of any reckless leading, I left without a trace of self-inflicted guilt. The guys sensing something was up were waiting with ears pricked forward like terriers. I sighed, my debriefings were rowdy enough without me having to describe to a bunch of thugs the problems facing medium-weight turkeys these days.

So ended a brilliant few weeks; it was just a shame that so many turkeys never made it to Christmas.

Several events in 1977 made a big impact on my future

Thanks to Barry Southwold's outstanding navigation skills we never stopped having fun on 208 and with his skills and my luck we won the annual Decci weapons competition against stiff opposition. Couldn't believe it. And best of all, Mike Bush told me over a beer that he was recommending me for the Buccaneer Attack Instructors course for a position in the squadron. I was stunned, to be honest not having yet finished my first Buccaneer tour. I said yes. Boss said yes. I bought the beers. Then Bill Aspinall called me in to see him – I wandered in wondering what rules I had broken – but it wasn't like that. He suggested I apply for the Red Arrows.

I decided to humour him.

I mean the Reds were right up there! Was I good enough? Did I need a haircut?

Ron Trinder was asked too and Bill told him the same. I could understand Ron being there because he was a really good pilot and never in as much trouble as me.

Bill laughed at me and said I might get better one day. That cracked me up but I applied anyway and forgot all about it.

But my application hadn't been forgotten. In the early Spring of '77 Richard Thomas phoned me out of the blue. He had been several courses ahead of me at Valley and was now number eight on the Reds about to take off for Honington and refuel for a display down the road; he wondered if I would like to fly in the back? I almost dropped the phone, hurtled to see the boss blurting out the Reds were coming in shortly would he mind me flying with them? And I met a smiling Richard on the pan after the Reds beat s**t out of the place in their red Gnats.

He introduced me to their Leader Frank Hoare a fellow Welshman who made me feel welcome. I recognised Dudley Carvell who had instructor me at Linton and exchanged pleasantries with them all as you do as a serf amongst shiny knights.

The whole thing was marvellous and lasted twenty minutes and I had no idea where we displayed! This relaxed atmosphere in this most disciplined environment was an eye-opener; there is more movement than you would think and slick formation changes under a lot of g were impressive. I didn't want it to stop. The beat-up when we got back was beautifully low – several feet high in fact – grass-parting wonderful and we looked about level with a tall thistle. I couldn't thank Frank and Richard enough.

Strolling to our crew room Richard asked me casually whether I thought I could do it. My answer came without thought – Yes!

Any problems like being up to my ears with masses of red metal bobbing up and down close to the ground doing 300 plus knots pulling lots of g never entered into it. Neither did the fact that the legions of hopefuls trying to join the Reds was impressively long. But hell! Think big. I wish I retained the same careless optimism for the QWI course. Mike gave me homework and had a lot to say about my untidy board work on briefings. I thought he was a bit picky to be frank. But when he remarked that some joyless hair-shirted nit-pickers on the staff thought that way: I concentrated on being a right Picasso.

But not to be! Someone from 16 Squadron took my place at the last minute.

Boss Rogers was sympathetic and said I would probably be on the next one if I kept my nose clean so I couldn't see any problem! Nothing could possibly go wrong.

chapter Thirty-Four

Shanghaied

Osprey launch HMS Sussex. Photo courtesy of Fleet Air Arm of Australia.
The RN Pre-Carrier course would teach us many things.

AS AN aspiring weapons instructor, I faced Spring of '77 with the confidence of a star batsman facing an inept bowler in an unexceptional sports school.

The first ball was straight – an invitation to participate in Red Flag a thundering good active exercise was coming our way. This was in Nevada flying with heavy weapons against fighters and missile defences in realistic settings. The workup had to be electrifying and was probably why we did all that ultra-low flying.

The second ball was curved. I didn't see it coming. And it came with the speed of a half-spent cannonball – deceptively slow but still capable of wrecking something – It seemed the Navy were short of an experienced Buccaneer pilot for the next cruise. It was kept quiet which was a shame as I would have been more on guard. Don't get me wrong I always felt flying off a carrier was an attractive way to go to work. But perhaps after my QWI course.

809 Squadron were good company and I liked their attitude. Accomplished flyers, they were a law unto themselves and fired off a lot of weapons for one squadron.

Tony Ogilvy was their punchy Weapons Instructor and OC 809 Tony Morten was quietly spoken with a well-deserved reputation as a squadron commander. Tony led 809 with a light touch and was an excellent pilot and his Senior Pilot Frank Cox was enthusiastically Navy and everything aviation. Anyway, 12 Squadron had more experienced guys falling over themselves for a chance to get afloat: The Navy could take their pick. And so they did. But not from 12 Squadron.

It started innocuously. After a week's bonza flying, Boss and Pete Jones plied me with drinks at happy hour and Graham drifted in with a noisy Ken Evans, Fritz and Ken Norman. Frankly, I should have been suspicious when I had difficulty paying for a round – the problem was that being an ex-single seat man I can be ludicrously susceptible to flattery and I remember through a fuddled haze that Tony Morten OC 809 was also buying…

I don't recall going home but I remember finding my supper – an overdone charred steak nailed to the door. But things got worse when I pitched up on Monday to chalk up a brief. Boss asked what I was doing there.

'Got a four-ship.'

'But Wadi on Friday you told Tony you would like flying off *Ark Royal* and you're meeting him this morning.'

" 'Did I' "? " 'Am I?' "

'Yup, good of you helping out Wadi; we're going to miss you but take the day off to meet your new boss…'

I had been 'Shanghaied'! and I knew who had been behind it. But in a rare flicker of insight, I realised bouncing off a carrier could be a new and semi-reckless way of enjoying flying. My brain sifted a few facts.

Was it my fault? – yes it was!

Was my ego down around my ankles? – no it wasn't!

Would I look like one of those plonkers who create a storm over nothing and then gets mad when it rains? – yes I would!

'Looking forward to it Boss!' And I set off for the Navy Hanger to find out what I'd done to myself.

Tony Morten thanked me warmly for volunteering to crack on with the Postgraduate pre-carrier course in a few days – which I couldn't remember doing – and after a brief chat, I hurried back to 208 to enjoy what little time I had left. On my last flight, Barry and I filled our boots with first-run attacks from Scotland to The Wash ranges. Pulling up to a tanker towline we refuelled and had a pleasant time boot-legging any range we could find for a dive-bombing finale. I felt a bit choked shutting engines down for the last time together: Barry had been the consummate professional and anything I had achieved on such a fine squadron was entirely due to his professional attitude on Buccs.

Although feeling downcast at leaving 208 a streak of optimism told me there were good times ahead – Dick Aitken was returning to 809. Dick a young tall Scot was finishing his Buccaneer conversion in '75 when the boss of 809 Mike Bickley ejected in Puerto Rico. This left the squadron short of an Observer. Faster than a cheetah on steroids Dick beat everyone's application did a quick pre-carrier course and before the ink was dry in his logbook shipped out to fly with 809.

As well as being more intelligent and less scruffy than me Dick was up for anything and I felt he would be a great guy to crew up with so I immediately started the rumour we were going to.

This was *Ark Royal's* last commission so it was now or never to be a tail-hooker and a real bonus to fly Navy with guys who had served on many carriers. I might jest with my dark blue colleagues but I respected their proud traditions, their ethos and the sacrifices so many made. Losing friends and colleagues in my first RAF operational tour with Coastal hadn't been easy but it taught me that respect and humour do much to soften blows. And blows I knew would probably come in the Fleet Air Arm. Flying from the cramped decks of a carrier in seas that favour no one attracts a degree of danger. But like generations before me – I couldn't wait to start.

I was expecting a ribbing from experienced Naval crews on the course but it wasn't like that. We were Crab-heavy with two other

RAF pilots: Pete John and Keith Oliver plus an American Navy Exchange pilot Lieutenant Commander Paul Barnard.

Back seaters were American Navy Exchange Observer/ Weapons Officer Mike Cochrane plus Flight Lieutenants Dick Cullingworth and Bobby Anderson.

Now, all we had to do was get on together. Oh!... and avoid drowning in the escape 'Dunker' at HMS Vernon's submarine escape tank.

chapter Thirty-Five

Fly Navy

courtesy of FAAA

IT WAS to be an agreeable and enjoyable post-graduate pre-carrier course. I knew it when a bunch of friendly Navy instructors in the bar told me I was only there because they couldn't find anyone else and I found a letter addressed to *'Flight Lieutenant Wyndham Ward. Royal Navy'* welcoming me to loan service.

My fellow course-mates were a mixed bunch and I took a liking to all of them – they were all volunteers which had to be a good thing joining any navy.

(I would probably have volunteered anyway if someone had bothered to ask!)

I was in a mild quandary where I fitted into the picture. I was there for Navy items like deck landings and catapult launches and it had been made clear for my former experience. But would there be a problem here? Going straight to 809 at the bottom and being a 'Junior Joe' in the Navy I could get myself into trouble adopting 208's forward-going attitude. Probably best to try and be polite to all things Navy and be quietly aggressive; in other words, get stuck in pushing the limits until I got a mild rollicking at which point I knew when to stop. Strangely enough, it started on my first trip.

Pre-carrier started with Lt Donovan from the OCU in the back with me leading a three-ship to the ranges. Pitching up out of the blue I hadn't a clue about anyone's background for they had been through the conversion course together and knew each other. To be honest they were a bit polite. And quiet. For years I couldn't recall anyone being quiet and polite at my briefings, so it told me a lot about their time on the course.

The problem of turning up at the last minute is that you forget you're a student again but the staff were brilliant about all that. They arranged formation positions so I merely put my stamp on it, briefed fuel calls and frequencies and went for a relaxed range brief: the weather was reasonable they knew what the wind was doing I wasn't wiping any bottoms go get good scores. Lowest buys the beer. Any questions?

A new boy on a serious course implies good behaviour. Ours lasted right up to our fuel check clearing the range. We had good scores, and stacks of fuel so I couldn't resist calling the boys into handy battle formation and briefing we were now ad-hoc bounce. Livening up life for formations returning to Honington proved excellent training and we got to know each other over a relaxed if noisy debrief. Apart from being reminded very firmly that our melees were completely unauthorised, I thought it a good start.

Next up, a three-ship for strike progression with no bounce followed by a break into the circuit at Honington for MADDLS – Mirror-Assisted Dummy Deck Landing sight. The projector sight was situated to the left of the *Ark Royal's* deck painted on the runway and a Landing Signals Officer (LSO) plugged into our frequency and stood next to it with a 'writer' who made notes while he gave advice. He could see the speed lights on our nose which indicated if we were bang on speed or not while looking through a sight aimed at us to check we were on the slope. If we were on glideslope he called 'Roger' and advised about power if not. And to see what we were made of, Jack Frost the fleet LSO arrived in a Hunter.

Jack briefed us on the projector sight then handed over to me and disappeared to sort his kit out. After briefing a navigation exercise to use fuel and a bit of strike progression I checked the aircraft and joy – when the aircraft numbers were handed to me I found a Sidewinder fitted to mine. OK a coal-burning mark 9B but I hadn't flown with a 'winder fitted since my conversion and with so much fuel it would have been a sin not to try it out. So I opted to change the brief on the spot. Just had to.

Standard aggressive stuff now. I told Paul number three 'no-frills battle'. Just cover my six, and I didn't care how he did it; if I wanted something I would call for it and two's be prepared to stick close so we could bounce any traffic we saw. Just relax, enjoy it, and don't be scared to take over and run the fight if you see an advantage. It was straight out of the rude Hunter handbook and did their faces tell a story? Grins, small nods to each other and a visible loosening up.

Paul was a natural number three and after some sharp handling from Pete and Keith who gave the impression they were having a whale of a time, I started cramping the inbound airspace to RAF Wattisham like old days – always good for catching a Lightning with his pants down low on fuel or even better a pair outbound with plenty of fuel who couldn't believe your cheek.

We ended up boxing in a Lightning returning home. During a hard turn to get a sidewinder growl (detection noise) it crossed my mind the Navy might not be as sporting about this kind of thing as 208. Anyway, it was far too late and my brief was to work the boys up so we carried on. Both Pete John and Keith Oliver had a discernible 'let's get stuck in' attitude which was brilliant because I suspected they had been given a hard run through conversion.

Buccaneer training was a gritty school with some hard nuts and I imagine graduate-entry guys would have been a target for the few instructors with a God complex. But it was different now. These boys were on their first tour going straight to sea like old Navy days and in my opinion a big ask – and perfect material to lead astray.

Paul Barnard the US exchange pilot and Mike Cochrane were hugely experienced with time in Vietnam; both were quietly confident and didn't miss a thing. Paul's polite unassuming manner hid a barrel load of skills and it was no surprise that he became the squadron LSO.

MADDL practice put us in the frame for deck-work and Jack went over various dangers involved with carrier approaches. Every landing is filmed so we viewed an impressive list of ramp strikes, explosions, wingtip collisions and aircraft going over the side. None of which put us off deck landings in the slightest!

Life would now consist of regular doses of sight work and for me: refreshing Lepus flare work for anti-fast-patrol boats followed by tactical work-ups for seriously bigger targets using Martell anti-ship missiles.

Lepus night attacks held a school-boy fascination for me with all its moving parts; plenty of pulling around and the flaming streak of rockets hurtling towards a target gives a positive feeling inside. Running in low and fast doing 500 plus pulling up to toss a flare and roll on your back to top out at 1200 -1500 feet, a quick roll to 120 degrees of bank and pulling round to fire as your flares silhouetted the target can sometimes be a bit disorientating. The unfortunate guy in the back for this was Flight Lieutenant Barry Chown.

Barry was called 'Wings' by everyone who in turn usually called people 'Shag' or 'Courtney' whatever their rank; and fun, trouble and expensive mess bills ran parallel to a good night with him – when tanked up on dining in nights Wings was a serial borrower of high powered motorbikes to hurtle around the mess. We got on well with much in common as ex-Signallers.

Next, catapult training. RAF Bedford had a shore catapult rig and we did three launches to qualify; acceleration was about 4 or 5g depending on wind and marvellous fun launching over a disused runway.

A carrier with attack aircraft has a CBGLO – carrier-based ground liaison officer – who talks aircraft onto targets during close-air-support operations. Major Anwyl Hughes of the South Wales Borderers Regiment was ours. Anwyl was a legend who didn't appear to have missed a punch up anywhere; he habitually wore a battered old jungle hat that saw service in the Malaya Emergency and every other campaign since. It had several patched-up bullet-sized holes in it and was as much 'Anwyl' as his Moses beard and enduring call-sign of Rubber-Duck. He was a popular presence in the wardroom particularly on formal occasions in his army mess kit of tight trousers and spurs, which he frequently entangled after a few beers.

The final piece for me was a joint maritime course at RAF Lossiemouth and all that was left was Divisional Officer training at HMS Excellent.

A Naval squadron is entirely self-supporting with cooks, stewards, aircrew and ground crews and the course was designed to help us deal with a division of Navy personnel. It was a seriously good course run well and helped enormously.

We were ready for the squadron.

chapter Thirty-Six

809 Naval Air Squadron

LIEUTENANT COMMANDER Tony Morten welcomed us on board. 809 squadron had a historic reputation for aggressive flying and Tony was a first-rate pilot himself who led with a light touch that was respected by everyone. Paul Barnard and Mike Cochrane have experienced many US Carrier landings but we RAF guys still had to prove ourselves by hooking on, and until we did there was always going to be a question mark in our minds.

The first thing was to fly a Bucc tanker. A refuelling pod could be loaded onto any of them and tanker flying was essential to carrier operations. Roger Carr affably gave me the lowdown on how it worked from the back seat and it was dead easy for me upfront. All I had to do was ask him to stream the hose and then fly smoothly. A tanker was always on alert to scramble and when it did another tanker immediately came to alert status.

It was possible to give away all the tanker pod fuel and yours which was another reason to crew up with Dick Aitken. Being a Scott he would not give away a fluid ounce more than he should because any over-enthusiastic giving away might mean diverting to a dry bar ashore.

For the first few weeks, we concentrated on anti-ship Martell attack formations. An attack by *Ark*'s aircraft was the original high-speed low-level penetration but these tactics could change at short notice of bad weather or a shifting scenario. Mike Callaghan flew with me to Spadeadam Range for a trial with our ARAM – Anti-Radar Martel missile. Mike, now on his third tour on the *ARK* had served on the carrier *HMS Eagle* and in his friendly calm style, he ran through tactics in simple terms and choices. He helped me to fit it all together in my mind for which I was grateful because anyone leading a large attack formation off a carrier has to think on their feet.

Pete John and Keith Oliver became even punchier with a cheerful standard well above their hours. When the main crews arrived back from leave I found them an adventurous bunch Bob Joy being the only squadron leader amongst us was our de fact unpaid Flag Officer Crabs (FOCRAB). He was a jovial dark-haired excitable chap and being an ex-manager of the Red Arrows Team could sometimes be diplomatic, which was good, because as FOCRAB he fielded awkward queries from the boss – mostly about the confusion over our uniforms.

All Navy uniforms were alike with white shirts, dark trousers, gold rankings and extremely smart. But the RAF wasn't like that at all. Half a dozen could turn up in casual items of uniform without anyone dressed remotely the same; I shuddered to think what FOCRAB would make of my tropical kit. I half-remember purchasing it in a beery bar auction from a well-oiled Gulf Hunter pilot; the only problem was he preferred Indian Air Force uniforms. Their khaki drill material was much cooler and lighter with a more elegant olive tinge than Gieves & Hawkes of London, and his Indian bush-jackets, knocked up in the local souk were far more dashing than the Queen's pattern. And a whole lot cheaper.

Naturally, the big question was when would we hook on for the first time? What was it like? Two RAF pilots Rick Phillips and Ed Wyer gave us the lowdown.

Rick was a likeable interesting character with aviation flowing in his veins; he had flown Buccaneers in Germany and his enthusiasm for flying off carriers was infectious. Ed Wyer, a tall lanky guy with a wicked sense of humour and great company was also ex- Germany and emphasised the screw-ups you could make.

Dates are firmed for deck practice in July. Couldn't wait and to cap it all Frank Cox our Senior Pilot confirmed Dick Aitken and I were going to crew together. It was the beginning of a professional partnership full of humour. We started by taking a four-ship against 56 Squadron phantoms for fighter affiliation; a great first sortie with six aircraft getting stuck in and Dick showing he was a born lead nav' and loved a scrap.

News trickled in for the Queens Silver Jubilee Spithead review; it's a Navy tradition and we were to do the 'R' bit of 'E R' for a flypast the Queen. Boss launched a nine-ship formation practice which was fun although the second practice didn't go as well. It started a ten-ship and became a nine-ship again when I managed to lose an 'awkward' engine.

I say awkward because that's how it felt flying formation with one good engine while the knackered one banged and belched away. After skidding around we eventually got bored with it and decided to give up and shut it down to comfortably make the bar before the others landed.

Ark Royal was ready for deck landing practice on the 4th of July. This involved flying to NAS Yeovilton in a Buccaneer and trying to be first in the queue to fly in a Hunter with Jack Frost for a 'look-see' at the deck; then have a go in a Bucc. Hooks up!

It's always been a privilege flying the Queen's expensive machinery but flying them at something bigger and more expensive like an aircraft carrier is something else. It's way up there; although I must say it didn't look that big when Jack talked me through the joining procedure. After a few approaches we haired back to Yeovilton for a coffee before the big act. Dave Thompson our Senior Observer (SOBs) was in the back with comforting advice which I fully intended to hoist on board. Only a lunatic turns the intercom off on his first carrier approach.

We made our slot abeam the stern running down the starboard side of Mother at 600 feet: this is 90 seconds before our Charlie time – hook on time. The idea is for me to peel off left downwind dropping gear and flaps to turn finals with airbrake three-quarter open to keep power up aiming for around 35 seconds between aircraft on touchdown. Flying accurately I will calmly call 'sight' the LSO would start advising calling 'Roger' when I am on glideslope. We will hit the deck firmly with no rounding-out and with our hook up, bolt off again with full power and airbrakes travelling in. Must say it all sounded so simple at the briefing.

What happened was slightly different. Having been told the key was to relax into the whole thing I wondered how the hell anyone could adopt a mental lotus pose throwing themselves for the first time at a steel deck. Turning downwind for the first go I remember thinking I was glad I didn't have to work for a living, all I had to do was to enjoy myself; how cool was that!

Dave's voice intruded politely into my relaxing thoughts. 'Nice. Keep that space out Wyn.' I acknowledge mindful that sitting behind someone like me on his first deck bashing needs nerves of steel. He was right, there's quite a bit going on here and a puny wide bomber circuit gets laughed at in the bar but on the other hand, you need a bit of time to get it right. Dave's quiet voice reminded me to fly *through* the wake before turning – the flight deck is angled off.

Thumbing the airbrakes three-quarter open and increasing power to bite into the drag a steady ADD note indicates speed smack on. A big ASI next to the strike sight confirms. You have to be accurate keeping 600 feet getting the turn onto the centre-line sorted and I manage to keep the height OK. I call 'three greens' and hope for the best!

'On sight' – 'Roger.' – 'Speed good.'

The meatball on the projector sight is nicely between green horizontal bar datum lights. The power of the engines is impressive and working against the airbrakes they give a fast response. Flight control corrections are crisp and instant because any approach from 600 feet on a 4-degree glide is sportingly fast. The line-up has to be good – any corrections closer in means eyes are off the meatball.

The dangerous bit of an approach lies in the 'hole'. Something you don't experience on a runway. The hole is a consequence of the ship being a large mass of metal superstructure steaming along 50 feet in the air producing a low-pressure area behind it. And when you hit it – you sink. And you *must* correct.

Murphy's Navy law of carrier landings puts this hole close in at the worse possible place. In fact, at precisely the point it's looking good and you take your eyes off the ball to get your act together to hit the deck. Dave reminds me in a firm tone when the meatball suddenly dips.

I shout 'f**k!'

LSO yells 'power!'

I pump it. We get high. I take power off. In seconds we hit the deck with over 1500 feet a minute down and a massive thump of squealing rubber prompts me to push two throttles fully forward and thumb the airbrakes in. Engines roar, a comforting massive kick in the back as the airbrakes smack in and power bites at the same time. Grey super-structure flashes past my right ear and in seconds the deck has gone. A lurch to the left as we pass the 'cow-catcher' bit at the end of the catapult and I gather the old girl check our speed and turn downwind at 600 feet throttling back – and start breathing again!

'Nice Wyn we would have trapped on that a bit high maybe but you corrected and a four-wire I reckon' Dave's calm voice informs me.

Unbelievable! My controlled teeth-juddering crash carried a hint of success.

'Slightly high and fast but I think I get it!' I croak calmly. No bull-shitting on a carrier approach!

Every pilot will agree: seen something once, things click and I started the downwind leg relishing another go. Dave was good at emphasising points just at the right time without distracting me and fair to say my concentration was different the second time and I notice a hell of a lot more. Line up sorted assertively: meatball checked every time I move a control by a whisker; any dip of the ball – power punched up and back and bang you hit. I did five 'DLP's and departed for Honington where I tried hopelessly to adopt a nonchalant pose in the bar.

The next day we did the same with Mother in the Western Approaches but this time Ken Mackenzie endures my agricultural flying. Returning home the ground crew inform me that I lost a

Penib fairing. I didn't know what that was but with my heavy bolters, I can't say I was surprised something eventually fell off. A shame because I was enjoying a good run of serviceability since my last engine failure in the ten-ship. Dick and I became well used to each other and up for anything which was just as well because we were told to report to the Vault Officer, who informed us we were to drop a 'shape'. A dummy nuke – and the brief was a corker: all we had to do was fly as low and fast as we could on a monitored drop on a laydown attack on Tain range in Scotland.

Grinning like a Cheshire cat Dick promptly knocked up a war-style low-level route around the north and I disappeared to find out what the maximum design speed for a Bucc was. I felt it might be wise to know what happened if you exceeded it.

No good asking a QFI; that would be a real bummer because they were bound to ask why and impose some rule or other. Instead, I sidled up to a few bold experienced guys and fell into shifty conversation. The census from Mike Bush was that if you exceeded 615 knots an aileron might fall off with 595 knots Indicated being a pretty normal limit. 614 was good enough for me.

A strict nuclear protocol is observed even for a dummy weapon and after signing for it we pitched up at our aircraft toting 9 mm Browning's in shoulder holsters. Squatting down to check our weapon we saw Navy markings. We expected RAF ones. But nobody had told us what colour a Navy drill round should be. Looking closer we saw the harness was RAF. The problem was I was certain I had been told RAF dummy harnesses were the same colour as a live RN one. Incredible really but on pooling our brain cells I thought I remembered being told this. It was one of those situations where you daren't ask for guidance. After all, you should know about this in the first place and a phone call expressing vague concerns about destroying a large bit of Scotland because I forgot what colour it should be wouldn't show us in a good light. As a crew, we made fast decisions so I simply selected low yield and winked, Dick winked back and off we flew.

The Range Officer's clearance crackled in my ears:

'Monitors running clear to drop.'

The airframe shuddered briefly as I flicked the weapons bay open and eased on more power. The turbulence seemed OK dropping below 50 feet and inching nicely past 610 knots I found the old girl steady as a rock. But the cross-trail was like nothing I had experienced – I was looking a tidy amount sideways. We must have been about 30 feet and I was feeling a bit sweaty-palmed to be honest and frankly concerned about my target way off left. But Dick casually mentioned it looked good and something about 'bloody big retard chutes' hence the angle.

Sight on, I squeezed the weapon off, powered back and easing up to 80 feet ran out at right angles (OK thing to do – the shock wave speeding from behind could flame our engines out) Dick counted down and as agreed we shut an eye apiece just in case someone had screwed up and the wee thing went off with a blinding flash and took half of Scotland with it. The monitor team thought it was a great run. They couldn't see us at first because we were lower than expected. On release when the chutes deployed they commented the weapon seemed to jump sideways and amazingly it walked up to the target so we got a direct hit. Dick's wind calculations had been brilliant and I had never flown so low and fast before in a Bucc. God, it was fun – and the ailerons never fell off! Just hoped our speed wasn't recorded because there would be hell to pay.

Embarking was confirmed as 5th September. It was a 'whole of squadron' experience where everybody from cooks to aircrew pitched in. A time too to hand out secondary duties to new arrivals. In the RAF to avoid this, you usually left the building early for home and hit the bar leaving career-minded chaps to soak up all the harder duties.

But that didn't work anymore. The Navy was different.

For starters, they were used to being afloat with no home to go to and for centuries they had handed out secondary duties in a strikingly different manner to us – every officer got a big one. Ever optimistic I kept a low profile with high hopes of avoiding anything strenuous but ultimately surprised myself by wanting the complete Navy experience and it wasn't long coming.

A vital part of a noisy wardroom is a piano. I had no idea what happened to the last one but 809 was supplying *Ark* with one from Honington. The piano was to be dismantled, transported and lowered by block and tackle in the Navy manner and manhandled into the wardroom where it would stay until fired off the waist catapult. With all the moving parts involved in getting it there, not to mention re-tuning, it was a predictable pain in the ass for anyone to supervise so when our XO asked if I could play the piano? I wasn't fooled:

'Tone-deaf Sir'.

'In that case Wyndham we have a Division for you!'

Taking on a Division of Navy men was one of the best things I ever did as an RAF officer.

August went by in a flash; we got as much practice as we could culminating in a big exercise during which I lost my main instruments during a toss manoeuvre. Probably not the best place for it to happen but for minor crisis moments like this Blackburn thoughtfully provides a standby horizon – it's a small cheap-looking piece of kit that almost forces a change of underwear even thinking about trusting it. But it worked.

I unwisely aired my doubts to an Air Engineering Artificer whose 'no-worries' reply was priceless. It was a question of Murphy's Law and basic entomology. Murph's law ensured the more complicated main instruments went tits up at the worse moment; and the cheapo standby instrument being smaller had far fewer bugs to iron out: so it was more reliable.

chapter Thirty-Seven

"Down hooks"

image © Mike Turner

ON 1st September 1977, *HMS Ark Royal* sailed from Devonport for her final commission. Her sister ship *HMS Eagle* is at anchor awaiting the breakers yard, she will provide spares for *Ark Royal* on her final West Atlantic deployment. Setting course on a fine calm day for passage off the West Coast of Ireland, *Ark* prepared for her Air Group embarkation on the 5th.

The Carrier Air Group (CAG) will comprise 14 Buccaneers of 809 Squadron. 12 Phantoms of 892 Squadron, five Gannets of 849 Squadron, seven Sea King helos of 824 Squadron and two Wessex helos for search and rescue. It is a heavy attack force with an equally heavy defence force supported by anti-submarine helicopters and early warning aircraft. On the morning of the 5th, ammunition and fuel are on board for the aircraft of the CAG.

And in the wardroom, one slightly dented more or less re-tuned piano awaits the boisterous arrival of the aviators.

There is potent anticipation to one's first hook on and I am sure Pete John and Keith Oliver felt the same. Dave Thompson sitting behind me must be feeling twitchy. As the senior observer, he is high in the pecking order and needs to be on board – I idly wonder if this is a compliment to me. Does he think I will make it on without tribulation? – or, does the boss thinks I need more supervision than the others? Possibly the latter. With a calm sea, a reasonable swell and winds not strong enough to give any trouble; I turned downwind with no chance of blaming the elements for any poor performance. Our hook is up.

Our briefing was thorough, casually given in the manner of experienced hands familiar with these things and we all knew what to do – a good feeling. In FlyCo, the bridge over-looking the flight deck, Little F: Pete Sheppard and Wings our Commander Air are both watching us carefully.

The Captain *of Ark Royal* Ted Anson is an experienced Buccaneer pilot himself and will be on the bridge to see his Air Group embark; and higher up on the goofers gallery grinning spectators crowd to watch the entertainment.

No pressure!

After three reasonable approaches the order 'down hooks' crackle in my ears. Wings think we can hack it! Must have done something right. I gave my harness an extra tug – going to need it – flying downwind I see a Bucc in the wires and it's my turn now: turning finals I keep the height accurate:

'Four greens' I call.

'On sight.'

'Roger.'

A steady audio note confirms on speed, the line up's smack on and as I peg the meatball with a prod of power I look forward: line-up still good and the wheels hit steel with a huge thump and I instantly go to full power thumbing airbrakes in – But a massive force intervenes.

No doubting when you trap a wire. Arrest is instantaneous. Flying speed down to zero in less than two seconds is one of the most violent forces you can ever experience without the thing coming apart. Flung forward into my straps, they dig in; my eyes tumble and I grin like mad. Dragging power off, the wire pulls me back a few feet and I select the hook up while thumbing airbrakes fully out. The wire falls away. It's the number three wire I find later. Marvellous!

No time to think – throttles back, hook and flaps up airbrake fully out, wings folding. The nose steering button engages and a marshaller pops into view with hands moving insistently. He wears a yellow vest and is leaning hard into the wind. I get handed on with a flourish to another yellow vest who takes me on and I am heading to Fly 1 in the bows to a mass of parked aircraft. My first impressions are of taxying uphill as the bow lifts and I need power; the deck is in constant motion and Dave gives me all kinds of verbal assistance. Behind me, another Bucc takes the wire and I concentrate on the guy's hands marshalling me in: they talk with their hands. I am impressed. Brakes on! Clenched fists tell me this. A hand crosses his throat: I cut engines. Seat made safe, chocks smack in and chains appear. Ladders wallops into the side, I lower my mask and climbing down I can't stop grinning. Despite being briefed about it the wind hits you at 30 knots and it's a shock.

Dave shakes my hand and keeps hold of it while I get my bearings with aircraft inches away; the deck tilts I stagger slightly: it could be the wind or the adrenalin but who the s**t cares! Engines blast, another Buccaneer strains against the wire, a good trap. I hope it's Pete or Keith! The noise is deafening, speeding grit stings my face; visor down I need to keep my helmet on until the safety of the grey Island.

There is an overwhelming physicality about everything going on around me: the noise has a raw potency and I refrain from punching the air. I get paid for this. Unbelievable!

Dave leads me towards the Island superstructure; we duck through a steel hatch and the noise abruptly diminishes. Outside, engines thunder as a jet misses the wire and bolts for another go, the whooshing powerful noise of its passage over the deck rattles the hatch. Sounds like I am on a London underground platform three feet away from a train doing a hundred miles an hour through the station. I sign the maintenance form still grinning...

We dump helmets and life-jackets and make our way to the briefing/debriefing rooms where I await Pete, Keith and Paul. Tomorrow's flying programme is up and Dick and I will launch for low nav and practice toss attacks. Marvellous.

Wings meets us making much of us being Crabs: how did we feel? That was good of him. I ignored the inner RAF voice telling me that if a national service Naval conscript pilot with terminal acne and a voice just breaking could do it – so could a Crab. But I was careful not to say that: Instead, with the biggest smile I could muster I said it was a privilege helping the Navy out!

An Aircrew Feeder below the briefing rooms supplies us with coffee and while we compare stories Dick wanders in with his usual grin and pokes me in the chest. He will show me where our cabin is. Mile Hall, Dick and I will share this cabin and I have no idea how to get there. It's cabin 6X which is way back aft involving a long trek up and down ladders through hatchways and along open walkways – flats in Navy speak. I'm curious and gawky as hell about everything. Dick gives a running commentary and will show me other routes in case this way is blocked by welders or working parties of fish-heads. It's a far cry from cycling around an RAF station!

Lower down, white-topped waves surging down the steel hull tilt the wet walkway under my feet and I noticed sailors everywhere; their numbers will swell noticeably when hands are piped down from flying stations. Six decks below, Dick leads us down a final steel companionway and I'm home! 6X 1. Keith and Pete will be in 6Y something.

Our sea trunks await unpacking and we get cracking. Dick has already claimed his bunk, Mike and I toss for the next best and I lose; Dick smartly retrieves his double-headed coin – both are navigators you see - and I sort myself out. Our furniture is comfortable high-sided bunks on top of a writing desk and drawers with a small storage cupboard. Luxury!

Tonight it's squadron dinner in the wardroom and we will take it in turn with 892 Phantom guys after a few drinks. The whole carrier air group will be three-deep at the bar. Our dress is Red-Sea rig which is a white shirt with epaulettes and uniform trousers with squadron cummerbund. Comfortable and cool.

What happened next was what the FAA was all about!

Life on a carrier without aircraft is a hollow existence. Cavernous hangars stand empty, hushed except for muted echoes from machinery above. No noisy pneumatic tools rip the silence apart; no bawdy shouts and ribald remarks signal any industry or repair – it's a barren place. And the wardroom's a desert to avoid. Until now!

Entering the crowded wardroom a wall of noise hits us. Two guys are playing loud jazz on the piano, a fiddler with a darting bow plays an Irish jig and Rick Phillips in full tartan gear plays the bagpipes at max' throttle. All play different tunes and everyone shouts to be heard. It's marvellous bedlam and all tension melts away with relief at making it on board; friendships are renewed insults hurled and life feels good. To my immense pleasure, I see Steve Riley my old chum from pilot training at Linton and RAF Valley. He is now flying Phantoms with 892; after all these years we are to be ship-mates and next to him is grinning Murdo Macleod also from Linton days, he is also on 892 which calls for many beers to catch up.

Despite our cabin sitting over four big vibrating screws propelling us along and with ears assaulted by noisy pumps, clanks and curious whooshes – I slept like a baby. The ship's motion rocked me to sleep.

Woken by a steward with a mug of strong sweet tea, a new day begins. It's early and dark and today Dick and I will launch for low level followed by toss attacks against a Royal Fleet Auxiliary (RFA) ship towing a splash target. It gives us something to do; the real aim is to get used to the deck. Pete and Keith are just as eager for our first catapult launch and with Dick in the lead we crowd through passageways, steel hatches and steel ladders past dark seas towards the ACRB the Aircrew Refreshment Buffet Here you get enormous egg bacon and sausages rolls washed down with big mugs of builders tea or strong coffee whenever the ships at flying stations.

Inside the island, our ready room is fitted out with rows of what looks like airline seats (economy class) where we join guys lounging around for briefing. Present are the Met Officer, Operations officers' Ops 1, 2, etc who give us current weather, air traffic and what's working or not on the ship, the fleet plot and any relevant information. This is where a formation briefs and Wings – Commander Northard notes everything. The ship's movement is of interest to me now with expected rolls and pitch. Our experienced guys have heard all this before but I'm all ears.

Long before my wake-up tea, the ships' company and air group crews have been hard at work. Shards of cold morning light break over a damp deck alive with movement; aircraft travel up in vast lifts to be towed aft and chained down by ground crews muffled against a thin cold wind who skilfully position them aft on a spray wet deck. They set the stage for us. We privileged few sip tea remote from the bustle knowing we will shortly earn our pay but whether or not we entertain the onlookers or displease Wings remains to be seen. My eyes are drawn to a list of pilots' names with a long row of small pins alongside them. Every landing is assessed by the LSO and he will put an appropriate coloured disc to a name. Our landings then become public knowledge and there's no dodging it.

Blue is a top-drawer landing and red is the other end of the scale.

The colour system is blue, smiley green, green, yellow and red. Last night in the bar we went through the colours where our unofficial explanation of the yellow end differs slightly from the LSO's...

Blue disc means a damned good stable approach, nothing is said by the LSO except 'roger' – on the slope with speed bang on.

Next down is 'smiley green', a green disk with a smile inked on it – not quite a blue but a really good safe approach.

Green – a good standard safe approach with maybe a speed call or two but nothing bad. Next is yellow subdivided into yellows 1 and 2.

Yellow 1 means a workmanlike untidy approach with multiple calls of speed and glideslope.

Yellow 2 disc has a frown on it and the LSO normally has one on his face when dishing it out. This is a shabby approach from start to finish with all the signs of an orangutan flying it. A late line-up, speed too fast too slow and then wrong and usually, you trap number 1 wire wondering how the hell you got there.

Red – is a barely controlled crash, provoking nightmares well past the event. In a rising crescendo, the LSO makes multiple calls about everything and you are barely on the glideslope at all – just long enough to hook on for a shambles of an arrest. The writer standing by the LSO has NEPICSAR written in shaky handwriting next to your number. This is LSO shorthand: Not-Enough-Power-In-Close-Sank-At Ramp. Just how you didn't get a ramp strike and fall apart is regarded as a miracle and draws applause from the goofers gallery. You are probably a bit shaky after all this, and after a one-sided debrief by the LSO the senior pilot usually has a few things to say before Wings does.

To be fair, apart from having a bad day the weather has a bigger influence on your approach than inept flying. In bad weather with a pitching deck, you will still be awarded a yellow. But it's no big deal and anyway you couldn't give a toss what colour you get because having scared yourself witless you are alive to get a rollicking.

Frank Cox Senior Pilot stands and quietly reinforces points on deck handling and we are good to go. Briefing over life-jackets on with helmets to hand we answer piped orders from the ship's broadcast.

'Man the Buccaneers - man the Phantoms.'

The voice is the rich baritone of 'Little F' Pete Sheppard. Dick dons his helmet I follow suit and we leave the grey Island making for 'Fly 3' astern where our Buccaneers are ranged opposite the Phantoms. Small tractors towing equipment criss-cross the deck and our eyes are everywhere. The Navy is good at presenting information with clear unmistakable hand signals and colour codes. Their colour system of tabards or surcoats - red for 'bomb-heads' white for handlers, yellow for marshallers and so on helps enormously.

External checks with the aircraft close to the edge are tricky: half of it hangs over the sea and I am aware of dark cold waves gushing past 50 feet below so I just count the number of folded wings look for a couple of engines and clamber up the ladder.

Our ground crews give me enormous confidence in the self-assured way they do everything and with our checks complete we watch the scene with interest.

Little F's baritone booms:

'Stand clear of intakes propellors and exhausts.'

'Start the Gannet.'

The Helo has manned and unchained and the Gannet will be first off and last to land. He taxies towards the waist catapult.

'Start the Buccaneers start the Phantoms.'

We fire up, engines rumble to life, the hydraulics give us brakes and I feel safer 'Buccaneers and Phantoms away chains.'

Figures dart beneath us the chains whip off – a thumbs up. We are ready and the flight deck becomes a wall of noise. It's a dangerous place with screaming engines, whirling rotors and the seductive mesmeric spin of propellers. Avoiding the powerful jet blast from bulky shapes manoeuvring in tight places is hard enough without the force of wind sweeping the deck and adding to the perils.

I know I am bound to get used to it all but these first impressions will stay with me for the rest of my life. Every player in it has to know what to do and has a place in the order of things.

Today my first lesson is to obey the marshaller. He's on a communication loop and will converse with his hands. And fingers. And head. A finger points straight between my eyes, hands sink low fingers twirl for me to slowly increase power; you read him instantly. I put on too much an immediate sharp shake of his head tells me so. We inch forward on a deck that's always moving and power has to be constantly changed as the bow or stern lifts and the deck moves. Breaking waves fling cold spray wetting bits of the deck to add to the fun. Suddenly the deck tilts sharply as *Ark* turns into the wind – our designated flying course. She can turn on a penny piece with her four big screws.

The marshaller leans further back as we turn into the wind; the rolling tilt angle seems improbable for such a big vessel; with a flourish he hands me to 'Y director' who will direct me to the waist or bow cat. He directs me towards the bow cat. Great! The bow cat is a third shorter and therefore the shot is harder because you have to reach the same end speed. I am now totally into flying Navy even before my first cat shot.

Behind me, Dick keeps up a calm running commentary; his eyes are everywhere and he is careful not to speak over FlyCo calls. My go. Taxiing onto the catapult is an accurate affair with a marshaller standing with legs astride the cat rail directing me; he occasionally uses a finger for finesse. I come up hard against hydraulic-operated roller chocks in the deck which aligns me accurately – at this point, you have to fight a pilot's usual impulse of putting the parking brake on! The FDO signals spread wing and behind us, seawater-cooled jet blast deflectors rise from the deck.

Ahead of me dramatic wisps of steam break away from the track where a massive heavy flat shuttle moving with controlled force speeds towards us. It slows to pass between the marshaller's legs and bumps under our nose wheel. If it touched his leg it would smash it.

White-jacketed handlers swarm under us checking the aircraft. Happy with our configuration the steel launch strop borne on strong shoulders is fitted around the shuttle and to hooks on either side of our belly.

A holdback is clamped to our rear under the fuselage. It has a frangible collar that will shatter and release us when the cat fires. A badger arcs his arm over in a clear exaggerated and unmistakable signal and the steel launch strop cable tensions immediately pulling us down onto our tail skid. Our nose wheel is high and clear of the deck in a correct flying attitude and it's a vulnerable feeling because if the catapult fires without the engines powered up we are in deep s**t.

The Flight Deck Officer (FDO) leans into the wind, his clothing flapping wildly; he holds a red flag behind his back and whirls a green flag for me to power up our engines to full power. Straining against the holdback the last thing I check is both wings are blown with the right pressures. If I don't like anything I will shake my head and he will slowly produce the red flag and the launch is cancelled. But we're good!

I accept the launch by raising my right hand flat to the starboard quarter light. The FDO sees it: I get an exaggerated nod and place my right hand on my thigh – the Buccaneer is a hands-off launch. My left hand is firmly pushing against the base of the throttles – if you grip them the acceleration will drag your hand back and you are in deep guano. The FDO glances back to

FlyCo the traffic lights are green and he sweeps his hand down – and nothing happens!

The cat operator has his hands on his head and doesn't do anything in a rush. Dropping his arms he hits the button and over a quarter ton of steam accelerates us down the cat. It's a short creaking ride and a bloody marvellous way to go to work.

No amount of briefing, not even the Bedford cat shot prepares you for the acceleration of your first cat launch from a carrier.

Accelerating through 140 knots in less than one and three-quarters seconds pushes our heads back pinning us to our seats. The aircraft is in trim for a quarter of a second after launch then our nose starts to rise. I blip nose-down trim, the gear's up, flaps travelling, and acceleration to 480 knots plus is brisk. Checks complete we head off for a spot of low level. At this particular moment in my life, I wouldn't want to be doing anything else.

Dick gets his hand in at low-level nav, Mother resumes her heading for a wide arc around Scotland towards the Moray Firth and we find our target for a bit of toss bombing. The systems are accurate and our scores are pretty good but will improve mere days into our sea time because the ground crew will fine-tune the kit.

Time to head back. Dick works out our slot time and with 15 miles to run we spot a pair of Phantoms breaking cloud heading in the same direction; I can't resist the temptation to bounce the beggers and close in hoping the Director on ATC has a sense of humour. He doesn't – I got told off and get sent to the wait. There is a low and a high wait. Naval terminology for holding patterns is orientated towards the DFC our Designated Flying Course. This will be into the wind and Mother could be steaming in any direction for tactical purposes and she will turn for our recovery at the last moment and increase speed. At launch and recovery stations just below 30 knots gives us good wind over the deck (a ship's engineer told me this worked out roughly three feet per gallon of furnace oil)

An alert tanker is spotted in Fly 2 near the Island ready to launch. The crew is strapped in bored to tears because they probably won't be used; at this stage, our brief is to return above max' hook on weight. This means jettisoning fuel to get down to it before the final approach – a Soviet frigate often brazenly sneaks into position to observe us and in later days I usually dumped fuel over it as a goodwill gesture. It's an inherited dirty Coastal Command habit of mine – makes them stop smoking, welding and using naked lights etc which I know is a bloody nuisance to them.

The Navy has a good reason for everything we do. Arrests into a wire or the hook bouncing and hitting a wire can fray it: so it's sensible to limit the weight that you trap it. If the cable parts under an arrest, you have a problem because with no flying speed left you will fall into the sea and sink. This is bad news. If you survive you might be off the flying programme for a whole day while they work out who is to blame. A Flight Deck Engineer (FDE) runs across the deck after a wire is re-tensioned checking for parted strands of wires. Four arrestor wires are numbered from stern to bow and the number three wire is the normal wire to trap so it gets the most use and if the FDE is not happy he will order it cut out. The landing sight will then be raised or lowered to target another wire. Number one wire is the last to be targeted being close to the round down as it invites a dangerous ramp strike if you screw up and break up. Apart from that, it's not a problem.

We have stacks of fuel, make our slot time and jettison down to our max hook on weight and I turn finals. I am edgier about my second trap because I know what's coming! And when it does I once again feel the astonishing power of the arrest – instantaneous. But this time my nose wheel is reversed after a longer pull-back. Dick mutters 'brakes' and I hit the right one and all's well. Hook up, select flaps up, fold wings, airbrakes out and power up fast with another guy 35 seconds behind us. Our marshaller hands me on and I shut down on Fly 1 with a huge grin I cannot get rid of!

chapter Thirty-Eight

Operational Work Up

EXACT POSITION Moray Firth embarked on HMS Ark Royal

The *Ark* and its carrier air group must pass an Operational Readiness Inspection (ORI) it's to be expected and the mood's upbeat. Getting to grips with a large division of guys at sea is hard work and Lt Cdr Tony Francis was proving a rockstar in helping me with the handover. Tony, a hugely experienced observer exuded reliability and steadiness and I was grateful for his abundant knowledge because it gave me time to concentrate on flying.

The Navy didn't lack talent at the sharp end. All the crews on 809 were a capable and forward-going bunch and all pilots were highly competent individuals. Tony Morten led from the front, Frank Cox senior pilot, Clive Morrel, Bob Joy, Ed Wyer and Rick Phillips were the old guard and Paul Barnard USA exchange and our new LSO was a veteran carrier man. This left us Naval Junior Joe's: Pete John, Keith Oliver and I to get proficient and providing we didn't screw up I intended to enjoy myself in the company of some seriously good flyers.

A purposeful air hangs over a squadron at sea. Everything we accomplished disembarked, we better when embarked. This is what the Navy did well and I felt swept up in its workings. Just about every sortie now involved air-to-air refuelling launching with max weapons plugging straight into a tanker streaming his hose ahead of the ship. Working up for non-diversion flying mid-ocean meant our refuelling skills had to be top-notch and on rare occasions, with a deck unsafe for recovery we could have several tankers in the air topping guys up. When the max hook on fuel is close to diversion fuel you get it right.

The whole concept of carrier operations calls for a robust engagement of skills and attitudes from everyone and I was attracted to its full-bodied intensity. The overwhelming physicality I noted on my first hook was still there but less daunting.

We kicked off with some forward air control work at Garvie Island just off Cape Wrath in northern Scotland. Garvie Island is about the size of an aircraft carrier and a realistic target for high-explosives. Mike Callaghan and I did a quick recce to Aberporth range for missile launch and I returned with Dick a few days later to fire one and got grumpy because the lead boys sunk the bloody target. Between Martell tactics, 8-degree dive-bombing and rocketing, we were getting there and on deck, Paul proved a remarkably good LSO accurate as hell about any below-par landings.

Surprisingly, it was a remark from our weapons leader that shaped my attitude toward future landings. From Tony's viewpoint, our job was sinking ships, destroying targets and creating a professional force, not brooding over bad landings. Adopting the fetal position just because we frightened ourselves fartless and got rollicked by the LSO was strictly for 'Jessies'.

Consequently, I never dwelled on my landings. Aircraft still in one piece when I walked away was fine by me and if we could use it again even better; as long as no complacency set in I was happy getting an adrenaline rush now and again if I screwed up.

On the second phase of the workup, His Royal Highness Prince Charles joined us in a Buccaneer flown by Fred De Labiliere. He thought hooking on was priceless and met us in the wardroom for informal drinks and dinner. At about this time we Crabs were growing beards which were at a particular scruffy stage which he found highly amusing. He had a phenomenal memory for faces and although we were naturally curious to meet him, nobody was in a hurry to crowd him at the bar in the evenings. When I asked if he would like to fly fast jets he replied wistfully 'I would give an arm to fly Buccaneer S2s with you all but it's against rules I'm afraid.'

Mike Hall and I felt this kind of impeccable taste and judgement is exactly what you would expect from our future sovereign. Everyone except the Phantom boys thought so too and it was a sad day when he left us.

The weather for his cat launch was good and HRH changing in the Captain's sea cabin discovered he had forgotten his hat. I was duty boy and took it to him and found him in his long john underwear grinning with obvious excitement which pleased me no end. Let's face it if you are getting promoted to King at some stage it's sensible getting your Buccaneer ride in first before you get barred by some royal rule or other.

Fred De Labiliere being a true Bucc' hand ripped past at 500 plus knots level with the flight deck, HRH went home high speed and we resumed flying.

Sliding sideways on the deck in bad weather with cat launches on a downslope into a heavy swell and generally getting wet crewing up became routine. For some pilots, the initial satisfaction in getting something new right gives way to a strong desire to experiment. Having heard someone in the bar talking guardedly about getting his gear up smartly I started selecting gear up myself in the middle of a cat launch. Meant fighting the g force a bit but I was assured it was safe, the 'weight on wheels' switch stopped them retracting – but it's not the kind of thing you ask the Senior Pilot about.

Boss gave me more four-ship leads which were brilliant. Launching off with a four-bore is quite different from a carrier because you have to get everyone together smartly as the order of launch is not the order of formation. Others could be launched before you and this reflects on fuel states. 'Zip lip' radio silence air to air tanking straight after launch made it harder to get your act together so the ability to read each other well helps no end. The emphasis was on flexibility with changeable weather which relied on a formation leader thinking on his feet. Hoped I was up to it.

Being cooped up onboard and then let loose in a remote low flying area brings on a strong 'end-of-term 'feeling. Pete, Keith and I found this out when I decided a good workout for armed recce would do nicely. The terrain is not mountainous in the extreme northeast of Scotland and things like double-decker buses tend to stand out as targets – even at 500 knots. It's a rugged place with not many people in it and the ones sitting on the top deck of the double-decker about level with Dick and me when we barrelled past looked a bit bored, to be honest.

Judging by the swerving around I saw in my rear-view mirror, I think the driver was taken by surprise again when Pete appeared even lower…

On debrief, to maintain future discipline, and allow for possible admonishment; by common consent, we kept mum about the Wick bus – we knew it was the Wick bus because Keith read the name on the front when he went down the other side of it.

Sadly there was some fuss – quite a lot really – and 12 Squadron got blamed. They protested loudly claiming they were down the South-West approaches off Cornwall at the time. Which wasn't much of an excuse and showed a complete lack of imagination!

The team for the ORI arrived in mid-October. They could task you with anything. As a new boy I expected to be wing-man to the experienced guys but instead found myself leading a four-ship armed with 540-pound bombs tanking after launch and toss-bombing live at Tain range. Immediately after recovery, a hard-faced inspection team tasked us with photo recce on a series of dams and we launched with a photo crate. Dick was smack on with his navigation and dusting off skills from past fighter recce' days I managed to describe the target without disgracing myself. Although I managed that a short time later.

Ark passed the ORI and Dick and I launched as a pair with Boss and SOBS – Senior Observer leading off the coast of Portugal for rockets and bombing at Allochet range. The weather deteriorated, the cloud base lowered and drizzle reduced visibility to marginal for weaponry. With accurate radio calls it's no problem for a pair and toiling joyfully away at rocketing and bombing I noticed an unusual amount of vegetation and scrubby bushes around the targets.

How the hell we managed to set fire to it when it was raining I will never know but a smouldering red glow in the scrub was producing lots of smoke. I called 'in hot' for a final pass heard nothing fired anyway and pulled off the target looking for the other aircraft But with thick drifting smoke and drizzle, I couldn't see him.

A knackered radio, bad visibility and an armed aircraft you cannot see are a recipe for disaster. Far better to go home.

Visibility improved but our radios still didn't work. Making blind calls we arrived at Mother's position. Nothing! This is not unusual and I was not going to interfere with Dick's navigation. He would be squinting at his radar checking his plot and as our fuel was about expected I relaxed enjoying the view wondering what ship we could buzz.

Dick steered us around but after checking out contacts for almost an hour the situation wasn't looking good. Precarious in fact!

Then I had a brilliant idea.

'Dick?' 'Yup' ' You busy?' 'Just a bit.'

'Let's make a creeping line ahead from the next contact, look for a bloody big passenger cruise ship, you mark it and when the fuel's about to run out we return and beat the hell out of it.'

'Aye and then what?'

'We bang out a mile ahead''

'Great! Make Captain's table for dinner?'

'Bound to invite us Dick that's our plan then.'

'Sounds good steer 045 degrees for a bit!''

With high hopes of a luxury cruise menu on the cards, we set off with not much fuel left. I spotted a large vessel about ten miles off but my heart sunk as our prospects for a good dinner were dashed. Starting a slow descent towards Mother our fuel state wasn't inspiring but I must say the ship's team were impressive.

I called 'Chicken fuel' which sets a strict protocol in motion. It's for a very low fuel state and based on the ship's safety, not ours. One attempt to hook on only. Miss, bolt or wave-off you *must* eject. No problems with that we didn't have enough fuel for a second go anyway! But FlyCo were professionals. I assessed the wind, jinked into it and sure enough, Mother immediately turned and we slotted in nicely. I decided not to play hero and use a single-engine flap setting in case an engine died on me – I only had one chance might as well give it a go with flaps I was used to, and strapped to the latest fashion in ejector seats: a double bang-out wasn't much of a problem anyway.

To be honest hooking on was something of a relief but it disappointed a goofers gallery packed with people who were expecting us to eject.

Wiggy Bennett and Twiggy Hanson, two lively and hugely experienced Navy pilots on 892 Phantoms assured me over a beer that minor dramas the first time embarked were to be expected and considered desperately good for a Crab's soul. These two characters had probably been down to chicken fuel more times than I had had hot dinners and it was reassuring to hear it didn't happen often. But they were wrong! Two days later we were even lower on fuel.

chapter Thirty-Nine

A French Incident

EXACT POSITION overhead Ile-Malone sixty-one kilometres south-west of Landivisiau

It was a fine day at 30,000 feet heading towards the Brest peninsula. Plugged into a Victor tanker our spirits were high and armed with anti-ship missiles we were nearing the end of our fuel top-up. Ahead lay the prospect of a high-level bash over France for a planned vertical plunge to low-level for a sneak attack on Mother defended by Phantoms who didn't know we were coming. French Mirage fighters were bound to join in provided the pilots had finished lunch, their radar worked and they could find us. A turning scrap at low level was on the cards.

I felt a muffled vibration behind me and ignored it. I was too wrapped up in the prospect of a good thrash around with Mirages and Phantoms to give it much thought.

Seconds later a pungent smell of fuel and a sharp call about a long trail of fuel behind us made me sit up. The sight in the rear-view mirror wasn't confidence-building; it looked like we had a stratus cloud nailed to our ass. Fuel was pouring out of us. Dick informed me that we were using fuel rapidly in the slow clear tones he uses when I've either fouled up or things aren't going well like a fire or something. That grabbed me.

We were a flying bomb. And that's not a nice thought plugged into a tanker full of fuel. We needed to get clear in case we caught fire or something dramatic and took it with us. But you never do things in a rush when tanking and buggered if I was going to make a radio call and risk a spark. It would probably have been OK but I was feeling cautious, to be honest.

Easing out we dropped slowly away and when clear I just said, 'Landivisiau?' – 'Yup'! Stuffing the nose down we whizzed past the tanker's starboard side so he could see us go and headed for the French Navy base. I could see the runways. Piece of cake!

The fuel smell wasn't good and Dick calmly informed me we had lost a quarter of our fuel. That's quite a lot to lose in a few minutes so I throttled back to get a decent dive going; and although I was wary of selecting anything that could cause a spark I tentatively selected full airbrake. We were still in one piece after I felt them bite but I told Dick to eject anyway if he felt uncomfortable. This was a bloody silly thing to say considering the amount of raw fuel swilling around; his rocket pack would have set fire to us, singed both our asses and make sitting down uncomfortable for weeks. Better to wait until both engines flamed out which by my calculations would be quite soon.

It was a fast descent and my first time under 15 miles long-final at 25,000 feet for a straight-in approach. But no problems we were booming almost vertically down with the option of a turn. Dick found a tower frequency and I punched out a Mayday call on guard before switching.

The French were a bit taken by surprise. When Dick informed me we now had less than a quarter of fuel *left*: I changed *my 'Bon Jour' to a 'Mayday'- avec un grande fuel leak, les probablement un ejection on les finals et deux chaps on-board s'il vous plait'* A strong Gallic silence followed. But I was going to land anyway whatever he thought of my French and somehow it worked out. I elected 15. 10.10 flaps unblown for a single-engine- fumes-only landing and by monstrous good luck, we made it!

A bit fast over the piano keys and wary about hot brakes with fuel leaking all over them I used all the runway. A leak big enough to lose our fuel in 60 kilometres had to be pretty large so our parking on a convenient bit was swift and Dick and I slid to the ground quite relieved.

Our arrival was pure theatre. A black Citroen staff car battered its way through tussocks of long grass bumped onto the runway and screeched to a halt. A lanky, painfully thin dark-haired French Naval Lieutenant swiftly unfolded from it and all limbs and grins he lunged for the rear door opening it with a snappy salute.

A short burly hairy officer in a French Captains uniform sprang out like a silverback gorilla emerging from a bush. Stalking up to us without as much as a bonjour he shouted, 'Why you land here?

Why you not eject? In French Navy, we eject when we say we do!' He mimed pulling an ejector seat handle several times in case Dick and I had forgotten how they worked but his excellent advice was interrupted by the blare of sirens announcing the arrival of a fire truck. The crew smelling the rich aroma of fuel stubbed out their cigarettes and sauntered over to watch the pantomime. Suspecting we had unearthed a rich vein of historical ant-British Naval enmity that no amount of diplomatic explanation could overcome: I gestured dramatically at our aircraft. It was standing in a smelly puddle with the last of our fuel dripping from the weapons bay – 'Voila!'

Buggered if I was going to say anything – I was trying not to laugh. Standing behind the un-friendly base captain was the lanky lieutenant pointing at his head grinning cheerfully and miming the international sign for a mad person.

We got the message. I promised to eject next time and not block his runway and politely asked if I could use a phone or a radio to call Mother. I said she was bound to be worried, but something got lost in translation because he threw his arms in the air and stormed back to his car muttering in French.

It struck me that perhaps the French pilots don't call their carriers 'Mother'...

The navy displayed their super quick recovery technique flying in a small team who knew exactly what they were looking for. A part of the main fuel pipe had come loose apparently and the syphoning effect did the rest.

Back on board ignoring gross comments about us wasting fuel we got stuck into exercising with eleven ships from eight countries. Mike Layard, our new Wings proved a hard taskmaster with high standards. The truism here is that 'a tough bunch needs a tough boss,' manning up for dawn launches was getting earlier and earlier. We had a small select team who were our night flyers and Dick and I weren't on it but it was beginning to look like we were. I decided to bone up on night procedures when I noticed fellow flyers disappearing completely into the gloom when manning up for a dawn launch and the marshallers used wands. Despite the red floodlights over the deck, it's still a shadowy place with dark corners and big huddled shapes looming out of the darkness as you walk towards them.

Loaded onto the catapult we could hardly see the end of it waiting for a sliver of light to split the horizon and make it dawn. Launching in foul weather hardened us up; the experienced hands may have seen it all before but I noticed a few raised eyebrows. When launching in a heavy sea if the deck is pitching the Deck Launch Officer (DLO) drops his flag when the bow points downwards deep into the bottom of the swell. And by the time the PO on the launch button has reacted and the cat fires the bow is on the way up and you hurtle off it missing the wave tops. Usually!

A pitching deck makes landings interesting. The armoured flight deck has small bumps and shallow indentations and the hook will sometimes bounce off them causing you to miss a wire and bolt. It's no sweat but a wave-off due to a pitching deck *can* catch you out. On Buccaneers with our blown wings, the procedure is to go to full power, airbrakes in, and *HOLD* the attitude. The increased flow over the wings has an immediate effect lifting you upwards and once well clear it's ok to rotate carefully. On no account do you ease the stick back and rotate while applying full-throttle on a wave-off! This places the hook lower down, you trap a wire airborne, the deck falls away and you lose all speed at 12 feet in the air. Not pretty!

A 3-degree true gyro-stabilised glideslope at one mile puts you at 300 feet. If the deck pitches 3 degrees or more you look level or lower than the deck a mile out - and going down! It's entertaining as much as it's alarming.

True to say Mike Layard our Wings was hugely respected by all of us; I might be new to the game but sensed he had that quality as a leader that made you want to meet his high standards even if we felt a bit twitchy at times.

Passage through Biscay and Gibraltar to operate in the Med' wasn't a time to put your feet up. Passage time had components. It was a boon for Divisional Officers calling men in for a chat and working out what's best for their careers. A time to cover aircraft emergencies, contingencies and target recognition; including interesting talks from Martell Liason engineers. I hadn't realised the French were good at anything except making wine but it seemed they regularly threw away obsolete circuitry and tracking solutions that some Brits and US engineers hadn't even discovered yet.

The Med was a launch area for mounting nuclear strikes with Buccaneers so it's a strategically important place for us to exercise. We could penetrate deep into Soviet territory, and knowing this, their planners answered with Su-24 and Tu-22 Backfire bombers. They were supersonic and hundreds were stationed in southern Soviet states and more in southern European communist countries. What made them so formidable was the anti-shipping missiles they carried. The AS4 Kitchen was huge at 38 feet long and the Tu-22 Backfire could ripple off three from a standoff range of over 300 miles. It was a fast missile with a speed above Mach 4 so the bombers needed to be destroyed before they fired. This was 892 Phantom's job. They gave our Buccaneers time to launch a nuclear strike into Soviet territory – or sink a Soviet task force – or both. They defend. We attack.

Exercises were enjoyably serious affairs. Dick and I frequently flew long-range low-level solo zip lip probe sorties with ARAM missiles to detect opposition radars. Some needed air refuelling on the way out and a quick turn into the wind by *Ark* for an un-announced recovery and minutes later we stood in front of the Admiral with our intelligence. It wasn't unusual to launch straight afterwards tanking up with a four-ship strike to the same target.

We developed enormous confidence in our ground crews who did some amazing fine-tuning work on the weapons systems, always a pleasure chatting to them in the hangers after flying stations which I suppose is the magic of a good squadron.

In December we steamed into Naples for a break. The city, capital of Compania was the birthplace of wood-fired pizza so a must-visit place for pizza connoisseurs. It was a good run ashore with never-ending red carafes of wine and tasty pizza when I had another brilliant idea. Why not volunteer to be the ship's expedition officer?

As far as I knew there wasn't one. To test this out Pete John and I took a couple of tents from the vast stores and drew victuals for a mini-exped. The system worked. We hitch-hiked a lift in Anwyl's landrover, climbed Vesuvius, ate a packed lunch and walked down to Pompei to camp with everything paid for. After several bottles of wine it seemed a good idea to get our hands on Anwyl's army landrover – and so began our plans for a trek down the Appalachian trail in the USA.

Steaming through Gibraltar, we turned for the Bay of Biscay to launch for Honington on 12th December. With just weeks to Christmas, Dick was determined we were going to launch. Loaded on the cat at full throttle, I was checking everything when out of the corner of my eye I saw the DLO sweeping his flag down – Dick had held a stuffed flying glove on the end of a long stick against my quarter light – we launched with my head sideways which gave me a cricked neck for the whole flight.

United with family was a time for reflection. I had reached an understanding of the 'esprit de corps' at the heart of a Navy squadron by having the good fortune to be on a happy ship with a great boss in the company of excellent flyers.

Clive Morrel, Tony Francis, Ken MacKenzie, Mike Callaghan. Tim Howard-Jones and Tony Ogilvy; to name but a few Navy guys always provided me with advice whenever I needed it.

All I had to do was enjoy the rest of my time with the Navy.

Flying when disembarked is surprisingly intensive because it's a good opportunity for renewing flying qualifications, heavy maintenance and air tests.

I flogged around in a Hunter with Bob Joy to renew my instrument rating and kept my hand in on night sorties. Boss and Senior Pilot encouraged me to work up Pete John and Keith Oliver which was good news because both were turning in solid performances as good as any of the more experienced guys with several tours under their belt. They were displaying cracking trends towards the cooperative belligerence that I felt defined Buccaneer operations – I suspected they may have picked up a few bad habits from me but that can sometimes be a good thing. Like learning to use a polite tone with a bristling range officer.

From single-seat days I knew they disliked range cowboys but sometimes wrongly think you might be one. In our case with full loads of 2-inch rockets to fire off on the Wash ranges,

I, unfortunately, picked up a sight depression fault. No sweat. Playing it by ear I asked for a 'best guess' depression of my own and salvoed a full load of four pods all over the range. It was pretty spectacular with a wall of rockets hurtling into the distance – most were in the range and a couple hit the target.

Instead of taking the formation home as bluntly ordered by a pissed-off range officer, I apologised for a minor sight problem promising not to fire anything else. I contended the others were OK and would be fine – it worked. They had good scores I bought the beers, the Avionics Pinkie split his sides at my best guess sight angle and I learned to apply deflection the right way!

Our serviceability record ashore was just short of excellent until Mike Hall and I air tested an engine, lost oil pressure and promptly shut it down. This impressed the ground crew no end because I hadn't wrecked it (I like to experiment to give a better picture which sometimes doesn't work out)

Good news came with the arrival of another new pilot, Scott Lidbetter. Scott was the last to train as a tail-hooker on Buccaneers and projected enormous enthusiasm. He also proved to be one of the liveliest and most pleasant characters in the Naval contingent and I felt we were lucky to have a strong showing in first-tour Bucc guys. Like Keith and Pete, Scott predictably threw himself into the flying and I liked his gutsy attitude. Practising self-illuminating night attacks using Lepus flares was a priority for me and my school-boy fascination resurfaced. It's more exciting leading a pair. Number two will track the target after you have lobbed a flair skywards from 500 feet doing 540 knots. The white-knuckle bit comes on over-banking to reef it hard around into a dive to fire off rockets before your flare fizzled out. God! I loved my job.

The second deployment passed in a blur and then came the big one. It's tough enough justifying to your better half that a nine-month West Atlantic Caribbean deployment is hard work. When she is several months into a pregnancy – it's trickier and marked the start of domestic shambles that lasted for the rest of my entire flying career.

Embarking on *Ark Royal* on her way to the Azores on the 6th of April 78 the wardroom was the noisiest ever and there was no doubting the crescendo of piano, bagpipes, fiddler and impromptu choir made leaving home for so long that much easier!

chapter Forty

The Atlantic Fleet Weapons Range

photo Ark Royal collection

POSITIONING OFF the Azores the Air Group picked up stragglers before sailing to disembark at Rhoosevelt Roads in Puerto Rico. The *Ark* is to berth there for three days and sail to use the huge Atlantic Fleet Weapons ranges of Vieques Island near Puerto Rico.

This break allowed periods with the ship's secretary as ship's expedition officer to sort out some remarkable requests.

Funding for a tour of chicken farms by enterprising ratings studying US farming was easy to turn down when I found out the chicken farms were brothels. Guys asking for funds to take gliding holidays in Florida got a closer look and I sympathised with the guys wanting me to pay for a dude ranch holiday. Ultimately I felt it was better to spread resources to both aircrew and ground crew and run a real expedition. Recruiting Anwyl, his Land-Rover plus trailer and his expertise was a doddle and before long we had a bunch of ratings eager to walk a section of the AppalachianTrail in the Blue Mountains. Pete John and I put our heads together with Tony Ogilvy and Anwyl and we quickly had a plan for a fine body of men who were up for it

Passage over, weapons loaded, the air group was ready to go. 892 Squadron is to live fire Sparrow and Sidewinders missiles at drones. 809 Squadron is to use high explosive 2-inch rockets and toss, laydown or dive-bomb with 1000 or 540 lb high explosive either direct action or VT fused airburst. If there were any sidewinders left we would fire them.

Naturally, it wasn't all play. We did some work. Or rather Dick did. De-magazining the ship down to our never-go-below level meant replacing old weapons with fresh ones. The Navy was into quality control so Dick toiled away recording the fuse and weapon serial numbers of whatever I threw around.

The choice of targets would swell the heart of any decent fighter crew. Forget millions of acres of blue sky to fire off Sparrows and Sidewinders missiles at puny drones – ours was big boys' munitions. Lines of airliners just waiting to be blown up or hosed down with high explosive rockets. Armoured vehicles in convoy lined the ridges like rusty ducks in a row or you could fire at isolated tanks with Anwyl 'Rubber Duck' Hughes talking you down.

In the matter of obliterating sea threats, the agile Septar fast boat target took some beating – armoured and radio-controlled this high-speed job was a beggar to hit because the operator saw the flame of your rocket launch and jinked it away.

To be honest I scarcely slept on the 21st of April. I was too wrung out tossing and turning in case an unexpected earthquake or tsunami cancelled our range slots.

But first, range familiarisation tossing practice bombs. I think the Range Safety Officer must have been new or lulled into a false sense of security with us Brits. Dick and I were the first to return with Pete John and Keith Oliver on the wing armed with wall-to-wall high-explosive VT fused bombs. These variable time fuses were proximity fused and set to explode at 60 feet. Lethal, they shower everything with supersonic shrapnel – and are extremely noisy.

Dick noted serial numbers. The RSO cleared us in 'hot'. I tipped in a 20-degree dive. We pickled off a bomb. Dick ticked off the number. The bomb exploded. Shrapnel ripped out listening posts. I got thrown off the range.

To be fair the RSO did shout 'what the f**k was *that*!' before we started to argue. As we Brits often used the place a measure of diplomacy was required, so I said it was a perfectly normal 1000 lb British round fused at 60 feet and what was the bloody problem? He questioned this saying it sounded much, much, bigger, so big it had blown his listening post away. *(I mean who needs a listening post on a range this size? You could use a tactical nuke on it and nobody would be any the wiser)*

Ever tactful, I sympathised, saying nobody likes losing a listening post but we had plenty more to drop and would appreciate another target preferably without any listening posts. With a choice to make he forgot about throwing me off, muttering something about 'hellova bang, typical nasty Brit stuff' and sorted out a target. It was a good start with only a minor difference of opinion.

This intensive weapons phase spread over April and into May. Launching at least twice a day with a minimum of four 'heavy' 1000 or 540-pound HE bombs or four full pods of rockets gave us an operational edge hard to beat. In one launch I fired off more than I would in years in the RAF. But no censure intended. It only points to budgeting because weapons are expensive – although I sensed the Navy didn't save for a rainy day they just blasted away.

The result of rigorous efforts on the giant Atlantic Fleet Weapons Range begins to show. Anwyl sets up shop accurately talking us onto targets, our orders are that no high-explosive to be brought back if it doesn't fire or drop off normally. Too dangerous hooking on so we resort to live jettison. But not in level flight. No way! Dick and the others were damned good at jettisoning live with amazing accuracy in both toss and dive attacks. Swiftly switching weapons and casually changing attacks can be entertaining provided you do it with bags of style!

Boss and I were having trouble hitting the Septar high-speed launch with rockets. The cheery radio-control operator was good at waiting for the flash of our rocket before jinking and our strikes were getting tantalisingly closer and closer – but no hits.

I thought I detected a jeer after Boss called off-target (I was fuming anyway and told Dick this lad needed sorting) Closing on the boss, I waited until he fired and with the operator distracted I called 'bombs' – we had two live bombs inside – rolled on my back flicked the weapons bay open called 'laydown' to Dick, sorted out a dive to 100 feet accelerated to 540 knots, confused the radio operator and pickled off two bombs.

The observer in the other Bucc claimed he saw two massive water spouts with the launch upside down in two pieces on one of them before it sank: I wondered if we had broken any rules but a jovial 'OK you win buddy!' sorted that out.

Tossing bombs around attracts spectators who shouldn't be there and it's amusing watching them fleeing like Pharaoh's chariots in the Red Sea. A nosey yacht sitting less than a mile out to sea off a target ridge with a naked lady sunning herself on the foredeck was a prime example.

Ed Wyer tossed an amazingly accurate bomb; I was downwind and observed it strike hard granite within inches of his target raising a huge puff of dust. Bouncing clean over the ridge it exploded out to sea raising a large column of water – beyond the yacht!

Leaving the industry and excitement of the Atlantic Fleet Range on 9th May, the *Ark* set course for the US Virgin Islands. Anchored off Charlotte Amalie I was unexpectedly troubled by minor feelings of guilt when I realised I'd forgotten to think of a name for a daughter. Charlotte sounded charming so I phoned the minute we were connected to a shore telephone earning myself a guilt-free run ashore.

chapter Forty-One

Working With Mother

Preparing for Flying Stations

IT'S AN eye-opener to participate in a large-scale exercise with the US Navy. If you are in the Jacksonville sea area with huge US naval bases in Florida and Virginia's backyard, exercises take on a whole new dimension. These are the home ports of nuclear carriers packed with mean aircraft like the F14 Tomcat so a lot of hardware is afloat in their parish. Couldn't wait to mix it with them and early May '78 through to September was a period to remember.

I flew my 1000th hour on Buccs, got mildly rollicked by an Admiral, joined those being told off for luring F14 Tomcat fighters down to 50 feet for a fight and flopped off the *Ark* with probably the slowest speed ever and somehow managed to get away with it. Oh! And the boys walked a section of the Appalachian trail and I got selected for a Red Arrows interview.

My bad day started well and a more promising start would be hard to find because Dick and I were briefed to probe for a major US target and get our finger out returning if we found anything. We could expect tasking with a four-ship attack on return and recovery was straightforward. Dick reported and swopped with another observer while I briefed quickly for another launch with missiles and a light fuel load to top up from a tanker. The technical log showed correct fuel, I signed up and manned up eager to go.

(just how the hell I ever managed to forget what happened next until prodded by Steve Riley many years later, I will never understand)

As fledgeling Navy pilots Mother obligingly steamed at max speed for us and the catapult put us above a safe 'Minimum Launch Speed' (MLS) by a healthy margin. But we were big boys now expecting a launch at MLS with just the right amount of steam for the catapult to get us there. This avoided Mother steaming flat out using a lot of furnace fuel – some wind is a bonus but today the wind was slack. The weight chit calculated for our lower weight gave the cat 'PO' the steam setting for our launch.

But at the end of the cat shot, we never reached the right speed.

I accepted launch and felt a kick in the backside with the usual creaking ride. But coming off the catapult nothing was usual anymore. And it was a bloody good job I selected my usual non-standard gear-up on the cat because I felt a vibration run through the airframe – I was stalling. Gathering the stick I eased it forward trying for more speed with the fifty-foot drop to the sea.

Bugger all increase. Definitely on the wrong side of the drag curve. The horizon was now unnaturally higher above my head and I expected my Observer to bang out any second. I felt for the 'Jesus Christ' switch – a shrouded jettison button to knicker everything off and call 'eject' while I sorted out the shambles. But using the outside horizon to judge body angle against the stall and not ditch was making it difficult for me to look inside poke around to find the button and fly as smoothly as I ever could.

God knows what the thump of his bang-seat would do but amazingly ground effect arrested our sink. But didn't do much for my speed! It all happened in seconds and it was all by feel now. Instinct told me our speed increase was painfully slow. A glance at the ASI was depressing but then the noise changed. Was that spray hitting my underside or not? Sod it. We were so low over the sea that a rebounding strike from stores bouncing off the waves would finish us so I didn't bother to jettison, I just called 'bit of a cold shot there' and hoped for the best.

I heard 'Launch cancelled' and luckily it all worked out, we didn't ditch and life looked much brighter.

It was obvious when we sank off the bows almost hitting the sea leaving lots of frothy wake behind that something was wrong. The guys would be busily re-checking and I went through my actions to see if I had screwed up. The answer wasn't long in coming! 'Check fuel.' – A 'roger' from me and a hearty 's**t' from my Observer summed it up!

'Max fuel?' I ventured.

'Sorry, Wyn I didn't check the tank indicator after seeing the tech-log fuel.'

My 'oops! max' fuel' call was something of a relief for flyco.

We had launched vastly overweight so everything was fine; nothing wrong with the catapult, launches re-started and my day improve but I wasn't looking forward to signing in.

The sign of a highly regarded squadron to me has always been the way something out of the ordinary is handled. Like fatalities or near losses. A calm look at the facts, no advantage seeking no opportunism and for this, you need a good boss and total honesty. Well, we had a good boss; he met me with his usual cheerful understatement, 'Gosh! Bit of a sink there I thought Wyn, good to see you back.'

And waiting at the tech sign-in desk was honesty in the shape of a familiar Naval Airman who looked nervous but determined to see me. As I ambled in happy at not getting wet – he stepped forward and said 'I'm bloody sorry sir, all my fault, I cocked up the fuel and could have killed you.' I recognised him as one of the best guys we had. Someone who gave me confidence in learning the ropes. No point in making a fuss. So I just said 'look it's my job to bring them back in one piece which I know I don't always manage!' It got a weak smile. 'Let's talk.'

Aware that a Captain's Table and a disciplinary cycle would kick in I knew it would be wrong to interfere with the process but on the other hand: a Navy squadron is closely knit and we rely on each other. I took him aside choosing my words carefully.

'If I'm any judge the one person who won't ever make that mistake again is you!' 'Definitely sir.'

'That's fine, you're probably going to get your ass kicked but afterwards just get back to doing your usual bloody good job eh!'

I put my views to Boss about the Airman giving me confidence on the flight deck with his competent manner and he nodded when I said I was prepared to speak on his behalf to his divisional officer. Small things but we all make mistakes and this was a classic case of a good guy trying to do too much.

Later that evening Dick Aitken and Tony Francis had a beer and a good laugh with me in the wardroom; Steve Riley couldn't resist telling me it was fun to watch and could I do it again for entertainment. Dick told me off for getting into more trouble without him and agreed with Tony's calculations that I was somewhere between minus 5 and minus 7 knots below the absolute minimum launch speed – not much in it really but good for the soul.

If Dick judged I was overambitious and used a US diversion with a good bar he adopted the custom of carrying spare jeans and a T-shirt for me when we diverted short of fuel. This invariably invited comments about times ashore but as a crew in heated debriefs we were modestly proud of our talent in dismantling egos; although I must say Dick could dish it out without a shred of diplomacy to any plonker who thought we were a pushover just because we were Crabs.

Flying got more exciting when the weather closed in. A Bucc crew in lousy weather will not fire off TV-guided missiles if the guy in the back cannot see anything to hit. The missiles transmit imagery back to the backseater via a data link so he can guide it. But in lousy visibility and low clouds, our answer to the maiden's prayer in lousy weather was four large bombs inside our weapon bays. Whoever led missile attack formations normally carried Martell ARAM missiles which homed to radar emissions so these plus our bombs gave us definite options to toss-bomb. The worse the weather, the harder we were to detect visually and during Exercise Solid Shield our target was USS John F Kennedy. She was big, mean and brim-full with Tomcat fighters.

The downside of modestly priding ourselves on being rather good at flying in low bad weather was that we inevitably had to fly in it. No point in rolling over and bleating we were outgunned either. Not the Royal Navy way. Nor this Crab's way either! When my name came up to lead an attack I decided to give it a full-on go. At the brief, the Met Officer pointed out we could expect the target area to have lousy weather with cloud down to the deck and miserable poor visibility. Not his exact words but that's what he meant. Paul Barnard my number three looked at me and winked. Great! The worse the weather the simpler the brief otherwise everyone gets confused – so I outlined the rough point I would waggle wings for them to slip into a 30-second line astern while still visual and emphasised – no clever stuff – it was 50 feet rad 'alt in places if required by weather to the target and zip lip. I would only blip a major change of course so just do it and keep to briefed speeds.

I knew Dick was astonishingly accurate in azimuth with our ARAM and we were fond of claiming flying at 50 or 80 feet was an OK thing; so let's prove it (Rick Phillips had the best description I ever heard of a Buccaneer in its element. 'Like a ball bearing rolling at high speed over a sheet of glass')

The most important issue in a rapid switch for a low-level in-trail strike is what you do after tossing bombs at 3 miles in poor vis' with fire control radar locking up. The best time to sort this out is at the briefing: I broke left the others alternating right-left-right from my break-out. This put Paul number three and deputy lead behind me placed for a join-up whenever he could. I had complete trust in his ability in radio silence, the same with Pete and Keith on the wing. Fanning out post-strike and diving back to 50 feet with a fast runout is exciting enough in ordinary bad weather; in the stuff we were expecting it was going to be truly sporting. 'Sort it out chaps' was all I said.

When I mentioned we would probe for the best attack heading so a tanker on the way home would be jolly useful, eyes swivelled to Wings. My remark of meeting us at a high level of say 200 feet got a grin from Boss and set the tone. Wings said yes to tankers and that was that! I like immediate decisions and the Navy was brilliant at this; operating from a carrier the facts and figures are all in one place and readily available to the cutting edge. Sure enough, the weather turned lousy I waggled on a suitable attack heading and the boys slipped into line astern while we could still see each other. It gets bunchy very low down in lousy visibility because you can't even see a bird until it whips past, but we clattered merrily on - and joy! Unbelievably we heard JFK on emergency guard frequency giving visibility less than one quarter statute mile and 'casevac' clearance for lift-off. Perfect! The casualty being evacuated in the chopper would be clear so I simulated launching all Martell ARAMs, and ran in for a brisk radar toss.

Things went wild on emergency guard frequency when JFK and escorts twigged what was going on. Comments got heated as they sometimes do with bad losers so I ignored them all. It was a full-on armpit sweating run back but as we cleared the worse weather I glimpsed Paul moving into battle formation and in my mirror, I saw Keith curving into position.

It's great working with professionals.

But that's not what the Admiral said to me in the ops room.

He raised an eyebrow when I said TV missiles were out so I decided to go for a pre-briefed bomb option.

'How bad was the weather?'

'Rubbish with poor vis and low cloud sir.'

'But you were below it surely?

'No, we were in it, sir.'

'What height?'

'About fifty feet sir.'

This may not have put us in a good light. But his remarks about 'umpires taking a dim view of questionable procedures in dubious weather with a live casevac involved' surprised me. Hell! It's the kind of thing we do. We don't obey a random set of rules thrown up by different bosses (OK some of us do but only the more 'bendable' ones) Any normal Buccaneer pilot would see nothing irresponsible about attacking a powerful carrier, blind, with the boys bowling along nicely behind.

However, I had learned much in the Navy and two things stood out. First, you never scratch the ship's paintwork with your aircraft; and secondly, a Crab does not give lip to an Admiral. I kept quiet.

The remark – caught 'em with their pants down and they are pathetic losers – never passed my lips. Why? – Because he was smiling. Which is the best time to leave the presence of any Admiral.

Ship personnel try to keep fit at sea. Some quite obsessively and when flying finishes the flight deck makes a good running track. Spare spaces below decks would be occupied by sweaty figures thumping hell out of hanging punch bags or grunting push-ups. And if space was enough to take a tiny mat it wasn't unusual to stumble across a tranquil figure in a full-on 'lotus position' meditating at full throttle oblivious to rattling pumps and riveting guns.

Navy chums told me the two hottest things to come out of Naval aviation were: 'The Races' and 'The SODS Opera' – Ships Opera & Drama Society: They were so right! A flat deck made a convenient racecourse and the huge lifts gave a stage which we all sat around peering down at performers. And gosh did they take it seriously.

Anyone punching bags or in serene lotus poses gets elbowed aside for SOD.s Opera practice. Not a racehorse to be seen though because it only leads to knobbling and underhand measures. Compulsory dress for the races was bizarre with a blend of the uncouth and the course itself was a work of art. Highly coloured lanes marked with numbered spaces, fences and water jumps with penalties formed a huge oval course. Propulsion was by two big dice and the mounts were toothy painted hobby horses ridden by jockeys in gaudy colours. Mike Hall was a natural for 809's horse. Dressed in tight long-johns tucked into big flying boots sporting a garish spotted shirt and flying helmet he entered the runners' enclosure dispensing black panther high fives to tumultuous applause – I lost five quid on him!

The Sods Opera, rated high on the international enjoyment index was fit for the London Palladium. Performances on the partially lowered aft lift underneath a balmy Caribbean night with the huddled shapes of aircraft looming over our shoulders were awesome. Today the whole thing would be banned outright!

The lumberjack song sung by a partially inebriated pilot in full Canadian Mounty uniform had major sexual adulterations to the script. But these were largely ignored because he accidentally stabbed himself with some scissors and we were all fascinated with the amount of blood splattering everywhere each time he dispensed lewd gestures during bawdy bits.

The band: 'Turk Thrust and the Y-fronts' who took their name from the only article they wore; were noisy but of outstanding quality. Their cracking line in deep pelvic thrusts in rhythm with the bass guitarist probably wouldn't go down well with today's censors and doubtlessly ensure a worldwide BBC ban at the very least.

Never before had I heard the fleet air arm take on 'Magic Roundabout' the Children's TV programme and I do not doubt in modern times I will ever hear it again. The gravitas exuded by three conservatively dressed gentlemen facing us over chapel lecterns was head-masterly. But the profanities and sexual allusions of a soaring imaginative nature drifting in dignified tones across a phosphorescent sea would nowadays ensure they remained locked up for a considerable time. Marvellous raunchy fun though.

A long time at sea for fast-jet aircrew who nature decided at birth to be risk-takers spawns certain competitiveness. Confrontations after a few beers start playfully enough by ripping someone's shirt pocket off. This quickly leads to wardroom rugby with aggressive punching – Officers don't fight. 'The Commander' who is Chief Executive Officer polices the wardroom, and the previous one didn't understand aircrews and lost control. Luckily we now had James Wetherall who knew all about aircrew. Immensely popular he qualified as genuinely brave by producing a referee's whistle and shoving himself into scrummages. Stooping over piles of bodies he usually blew before the first punch had been thrown and would yellow or red card an offender at the drop of a hat. We were lucky to have someone of his calibre.

It was time to disembark at Cecil Field Florida where we experienced a bad weather recovery US Navy style. Thunderstorms brew up quickly in Florida, they are powerful and we ended up dodging storm cells, booming thunder and sharp-edged downbursts. (It was real Hollywood stuff of the kind you get to see when John Wayne slaps his co-pilot's face between flashes of lightning and tells him off for being scared)

But I'm a Navy pilot now, and tradition demands I look suitably unconcerned in the face of darkening skies, forked lightning and aircraft all over the place running short of fuel. The situation gets mildly frenetic to which the US Navy's answer was elegantly simple. They assume their boys can handle crosswinds and tailwinds; in fact, any kind of wind and know how to take a wire. So we hooked on from opposite ends and ninety-degree crossing runways simultaneously and everyone was down in no time at all. Handling crews had us out of the wire as if we were on board. Sporting stuff and the way to go.

It was 22nd June and we had aircraft disembarked until August so a small group of us planned n a trip to Texas, California, San Francisco Bay and on to Fallon Nevada. We planned to go north to Washington but Pete and I as a pair never made it! Pete aborted take-off and as Fallon is very hot and high his brake temperatures went through the roof. The good news was he was smart enough to turn off the runway before his brakes welded on. The bad news was waiting for a ground party to fly in and fix us which meant missing the start of the Appalachian Trail expedition.

Pete and I returned to Cecil Field after beating up Fallon airfield. A US navy DC9 airliner flew us to Atlanta and we then hitch-hiked to the finish point aiming to backtrack and meet up. Every American we met exuded help and generosity which was quite humbling. To be honest I felt grumpy but that soon disappeared when we met Tony Ogilvy, Anwyl and the boys. They had done exactly the right thing by setting off and keeping to the plan and despite missing some good walking it was a good break for everyone involved.

Life was bowling along nicely when the boss told me the Red Arrows had invited me for interviews in the UK. Would I fly home immediately? I was tremendously pleased to hear Steve Riley of 892 phantoms was invited too.

The Red's high standard of flying discipline was something I admired greatly.

The Reds high standard of flying discipline was something I admired greatly

chapter Forty-Two

Crimson Crab Selection

image Arthur Gibson

IT'S A benchmark in anyone's career If the Red Arrows shortlist you and a call for celebration whether you get selected or not. Riles and I were soaking up our good fortune by getting pie-eyed before we left for the UK which wasn't unusual for us having bumbled through pilot training together. Ending up on the *Ark Royal* as cheerful apostles of Navy aviation had been a good twist of fate and we could scarcely believe that even better flying might be coming our way. Best of all there was nothing to do. Not for the selection anyway and after that who the hell cared; romping around doing things with aeroplanes that would normally see you court-martialled couldn't possibly be a grind. Neither of us was remotely inclined towards point-scoring on interviews; all we had in our favour was a passionate desire to fly with the best aerobatic team in the world and it was bound to show over a few beers – which is roughly what the selection procedure turned out to be.

A surprise visit to a wife during pregnancy mostly spent alone put me up on points. Basking on the moral high ground was something new for me and I hoped this would carry over into the interview and selection. The process consisted of aspiring pilots cramming into a local Cotswold pub and getting stuck into tankards of Cotswold ale with Frank Hoare and the boys. Frank, as team boss went out of his way to relax us. Richard Thomas was generous-spirited as ever and a smiling Bernie Scott popped up from Hunter days. Tim Curly and Steve Johnson I met for the first time, both had flown off *Ark* on Phantoms and were sympathetic about our long cruise

RAF Kemble home of the Reds has a homely feel and is not at all pretentious. The famous red Gnats lined up outside looking tiny, I knew some of those on selection, as did Steve but ignoring a nervous edge to the crew room we amused ourselves leafing through the team's private diaries.

I wasn't too sure about an interview with the Commandant of Central Flying School. He was a tall immaculate chap who looked as if he was into asking questions requiring perceptive answers. If that was the case he was out of luck with me. But amazingly apart from initial pleasantries I only said one word!

The Commandant turned out to be jovial and polite. His manner was a terrific blend of self-assuredness and a well-developed sense of pride in the RAF that men who are going places possess in abundance. Jet-lagged with a small hangover, I wasn't in the same category and half-wondered if he would ask my views on the defence budget just to see if I was more articulate than I appeared. Providing he didn't probe into anything to do with flying discipline I wasn't too worried but astonishingly he closed my file with an elegant gesture. For the next five minutes, he praised the efforts of past teams and set about the proposed introduction of the Hawk. The team's international reputation with a probable sales drive for the new trainer was great stuff. And all one way. Although I expected headwinds when he upped his tone emphasising that not only was the training tough and exacting but I would be expected to act as an ambassador for the RAF and country. He leaned forward.

'Are you up to it?'

I just said 'Yes' and I meant it right down to my boots. He didn't ask for my views so I didn't give any. Just held his gaze. A long pause made me wonder if I had overdone the 'silent type' but it was over one way or the other now. Encouragingly he give a relaxed grin then he shook my hand.

This is where we go home and the team takes a few days to put their feelings together. It's a very confidential process where nothing leaves the room – Ever.

Competition is fierce and any remarks discussions and votes are 'team confidential' for good reason. Some team pilots will be leaving some will have one or two more years to go and be friends of aspiring pilots and will naturally plug for them. Some candidates may not know anyone but have outstanding flying abilities and need a chance. Too many from one force or background will fracture trust so there are many factors at play; but one thing is clear – you have to get on as a team under huge pressure.

There are no prima-donnas on the Reds.

The team adjutant asked us to phone in before we left RAF Brize Norton for the Caribbean, he might have news! Steve phoned in and I crossed my fingers for us both. 'You're in I'm out' was all he said. He extended his hand and despite being selected I felt crushingly disappointed we weren't doing it together. But you cannot keep a good man down and I took away the thought Steve was younger and had accomplished much on the Lightning and Phantom force with time ahead for bigger things – which turned out to be true.

A changing future for me meant less time with the Fleet Air Arm. The last catapult launch was sometime in November and the Reds normally start winter training in September. Fortunately, two exercises cemented forever my connection with a rugged aircraft and the Fleet Air Arm – Exercises towards the Azores and a massive scale Northern Wedding out to 61North 20 West, lasting several weeks into September with non-diversion flying involving 200 ships from nine nations including USS Forrestal.

Both heavy jet squadrons were probably at their peak and aggressive flyers with it.

Phantom guys all had to get night qualified which made sense so they could launch at any time to do their stuff. On Buccaneers we had a small dedicated night team to press home night attacks and they had to be in the groove so the team was small to remain in practice. Boss and Senior Pilot informed me I was now the first reserve if anyone fell off for sickness etc. As one of the newer joiners, I was pleased and surprised to be selected. All I had to do was a few 'duskers' to get a taste and keep sharp until required. I was under no illusions about that having already experienced the Navy perception of dawn where I couldn't see the end of the catapult and wondered what dusk would be like! Sure enough, I could barely see the end of the catapult and who needs a horizon anyway.

After a launch and immediate transfer to instruments, there seemed no point in trying to keep Mother visual so I eyeballed it from a CCA –a Carrier Controlled Approach with a relaxed voice talking me down to three-quarters of a mile. Getting a 'lookup for sight' call I broke cloud cover – lined up on a dark hulk with startlingly few red lights to sort it out and followed the landing sight to hook on with a mighty thump. Challenging was putting it mildly; it seemed to the division between survival and suicide at night was a small enough margin without anything going wrong. I looked forward to the challenge.

Years after I left Pete John recounted to me what could go wrong. He had a twenty-degree compass error on a CCA at night with multiple corrections and ended up a mile out level with the deck. He also had a big hydraulic failure after landing when he taxied onto Fly One; his brakes failed and the only thing that stopped him from plunging into the sea smack in front of the *Ark* – which is not a nice place to be – was a small inches high metallic rectangle on the edge of the deck at fly one graveyard.

In friendly comparisons between RAF and Navy, the question of flight authorisation regularly crops up as a baiting issue in the bar. After listening for the umpteenth time that Crabs were wimps who had needed a nursery signature on the authorisation sheet before flying: I pointed out the Navy did so as well. 'Aahh! But, your *actual* authority is the green light on flyCo which the FDO checks before he drops his flag' was predictably trotted out by dark blue crowding the bar.

This remark proved untimely because like everyone else I was moving towards one of those peaks in performance you get when at sea for a longish period. Nobody gets complacent with deck operations and survives – but with honed-up skill levels, there is a definite cycle of measured risk-taking, some friendly rivalry, much un-friendly rivalry and cheerful bloody-mindedness. All pretty normal as far as I was concerned and I noticed it continued right up until you frightened yourself fartless. This then usually forced a reset of attitude back to a more gracious form of aggressiveness and being nice to people again. And so on.

Unfortunately, I was on my way up the cheerful bloody-mindedness slope when it coincided with Flag Officer Crabs having a grumpy morning. I was leading a four-ship on a dawn launch in challenging weather and he was authorising officer. Ops officers trotted out good intel' and the Met Officer's gloomy undertaker look shifted towards a weak smile when he indicated a patch of better weather an hour away.

I noted a tanker would be on alert status with another on standby for recovery; the Ops guy Nick Tobin was well ahead of the game and he had it all sorted so I couldn't see any problems. If it all turned to worms I would simply break the formation down to radar-carrier controlled approaches.

Good stuff. Go for launch. Do the job. Sweep up with tankers. Wings would have the big picture firmly in hand and would expect a launch from me.

But FOCRAB had other ideas. He folded his arms and refused to authorise me – 'weather'.

But! Halleluiah! What a super opportunity to explore Navy claims of 'sod the paperwork we only need a green light to launch'. Promptly changing my 'begger off' to FOCRAB into a 'let's go boys' I led them past the auth' sheets studiously ignoring the big blank space where the authorising' officer's signature should be.

Engines thundering, we shook and shuddered on the cat'. Leaning hard into the wind the FDO's green flag circled furiously above his head, he made a quick backwards look at FlyCo. A green light! His flag swept down. Dick cheered I yeehaaad! And two illegal Crabs hammered down the cat. Way to go!

A challenging join-up before a good hard sortie with impeccable tanking and a beaut recovery by Pete, Keith and Paul brought home to me what I would be missing when I left the squadron. And judging by the long list of execs and FOCRAB waiting for me when I amiably strolled in after hooking on: my leaving looked imminently.

But I was wrong. Boss was brilliant – he confirmed I was leaving when we put into Moray Firth in September and must have forgotten to tell me off. The Senior Pilot tried not to grin when I pointed out we had a green light so what was the problem? He said it was a matter of good form to be signed out and would I mind doing so in future? FOCRAB joined in the spirit with a grin and we 'horsed' for wine at dinner. I lost and had to foot the wine bill – which was probably as it should be for me being so bloody-minded and we still laugh about it whenever we meet. And best of all not one Navy guy mentioned anything about authorising signatures and green lights to me ever again!

Joining the Reds in September was getting close so Dick teamed up with Ed Wyer to keep him out of trouble on the night team and I flew with Bobby Anderson, which was predictably fun as we had both come a long way since our pre-carrier course. Dick and Ed diverted due to bad weather and a pitching deck as we went through the tail end of a hurricane. Taking advantage of diverting ashore, Dick checked with my wife and flashed a message over the defence net – a baby girl and both well! A daughter called Charlotte!

My last trip was a tanked hi-lo strike into Norway where Bobby and I had a ball flying as low and fast as we could and the boss kindly met me after I trapped for the last time. He was as good as ever at judging my feelings and could see I felt totally at a loss at leaving the squadron. The cheerful note was my last deck landing was a blue! What's more, the LSO was my good friend and colleague Paul Barnard, we shook hands to end a well-respected partnership together.

Time for reflection. Standing on the quarter-deck in Red Sea kit dressed for dinner and awaiting friends brings a relaxed feeling. If launches and recoveries have been spot on it's usually a fun dinner with flamboyant characters agreeably mocking each other. 892 characters like Wiggy Bennet, Twiggy Hanson, John-Eaton Jones, Riles, Murdo and Tim Hewlett were always fair game to bounce on recovery and good for banter – even if they got waxed – and if things haven't gone well you might as well share a bottle of wine and forget about it. Dick made a point of doing it whenever I screwed something up.

The sea breeze playing over my face draws me into thinking about the life I have led. The look-outs discretely leave you alone and beneath my feet, the deck vibrates gently to the idling power of four screws churning the sea into a white froth astern.

I know these muted vibrations are deceptive. They conceal the gutsy strength our carrier needs at flying stations when more power is called for. I reflect on the way the Ship's Company support the flyers to make it safe for us to play our part – always reliable, our belief in them leads to expectations which I suppose translates as trust at the highest level.

There are many levels to the tiers of trust on a carrier.

Faith in the abilities of the squadron guys flying with you is a given.

The men in your Division rely on you and you, in turn, trust them and their loyalty.

Naval Airmen who talk with their hands marshal you within inches of danger. Black and white-coated Badgers fussed around you with the handlers ignoring harsh biting winds and bitterly cold spray to keep you secure perched on the immense power of a steam catapult. Armourers produce an awesome array of weaponry from the ammo lifts and this heavy ordinance gets attached to your wings as if by magic.

Flight Deck Officers and Flight Deck Engineers make snap decisions to prevent a tragedy. Fire crews hover like watchful hawks. And hard-assed clearance divers onboard our SAR helo' are scared of nothing, ready to drop on you in times of distress – and I haven't even started on the medics, ship's surgeons who high in their fields exercise restraint as we revel in the wardroom!

Nobody cares what colour your skin is or what your religion is or for that matter where anyone bloody well comes from – providing you weren't a Crab of course.! Loyalty to your comrades and your ship matters most. Dick, Mike and I never fell out once in a crowded cabin. Early anxieties were dispelled because I was lucky to have good senior commanders as I had in the RAF. I think the hard nature of life at sea flushes out any weakness and therein lies the good fortune I have experienced with OC 809 Tony Morten, Mike Laird our Wings and *Ark*'s Captain Ted Anson (Mike Laird wasn't just an excellent Commander Flying, he flew in the back with me when we took on the whole squadron in a weapons competition in Florida and together we won it).

Tomorrow I will be choppered out from the Moray Firth and go south to meet my new boss Squadron Leader Brian Hoskins. We flew Hunters together and he was a Buccaneer pilot – another good boss!

Setting off to work from the *Ark* held a certain style – will miss that.

Photo Library Ark Royal collection Caption 'Setting off to work from the Ark Royal held style and panache which I shall miss'

chapter Forty-Three

The Red Arrows

'Back to Gnats'

THE GREY expanse of runway looked absurdly long after the deck of a carrier. The snug cockpit and the muffled growl of the Orpheus engine remind me of distant training days. I grin into my oxygen mask recalling how the little Gnat with its long pitot head poking way out front gave the impression of riding a witch's broomstick!

Two other guys probably thought the same; Malcolm Howell and Ray Thilthorpe were also converting to the Gnat here at RAF Valley to join the Reds.

Malcolm is an ex-Phantom pilot, a tall quietly confident guy with a modest air about him was just the sort of guy you expect to be on the Reds. He will be Red Four. Squadron Leader Ray Thilthorpe is the new team manager who recently flew Canberras and is converting to fly the spare jet as Red Ten. He will be the team commentator as well as the Manager so had a lot to get his head around. With a sociable attitude, good buzz lines and lively banter he exuded a worldly air that I imagined was indispensable seeing off the army of smoothies in the world of air displays (turned out he was well selected in that respect)

These fine gents were probably surprised when I pitched up. My decent suits were in transit anywhere north of Gibraltar and I hadn't had a haircut or seen a runway for months; so heaven knows what they thought of some scruff lounging in the bar claiming he was there to join the Reds!

But we hit it off and had ten flying hours to get used to the Gnat.

'Clear to go' from the tower. I opened the throttle. My brave instructor one Flt Lt Mullen murmured something: I didn't catch what he said and didn't trouble to ask because I was entranced once more with the playful energy of the Gnat.

Recent habits die hard and at unstick speed I whopped on the bank peeling off the runway heading to make room for another jet selecting gear and flap up all at the same time... Oops! I accepted the ticking off gracefully.

It must be difficult instructing young students hanging on to your every word and the next minute find yourself behind a docile ruffian with a breezy air about him. Bound to raise alarm bells. It's so easy to give the wrong impression by asking to pull the wings off in max rate turns before having a go at engines and hydraulic failures over a loop...

But the instructors were terrific and the only remark boss Mike Keane a popular New Zealander made was my touch and go's used less than ten yards of runway. In five days we converted and set off for RAF Kemble home of the Reds where we would meet with Neil Wharton the new Red Eight.

Brian Hoskins was our team leader and Boss, our adjutant Warrant Officer George Thorne was known as 'Uncle George' and the whole assemblage was commanded by Wing Commander Ernie Jones an ex-top notch Lightning display pilot known as Little Ern. A more friendly capable character you would be pushed to meet. The Team made us welcome with quiet enthusiasm which was the hallmark of the Reds. Any concerns we harboured at joining them and not letting the side down were short-lived:

the team pilots were modest individuals with a wicked sense of humour and knew exactly how we felt. They just wanted to know how we took coffee. OH! And would we be perfect in Diamond Nine by Tuesday week?

The team of 1979 was: Brian Hoskins Leader, Tim Curly Red 2, Bernie Scott 3, Malcolm Howell 4, Martin (Stumpy) Stoner 5, Richard Thomas 6, Steve Johnson 7, Neil Wharton 8 and myself 9. Normal pilots fly a three-year tour, changing three pilots a year and usually, a pilot will fly on the left or the right side of the formation and sticks to that side; the finesse involved and the perception of feel and movement demands this.

Life bowled along bewilderingly fast: having left the *Ark Royal* in late September; I met my new daughter Charlie for the first time, converted to the Gnat, joined the Reds and on 16th October sat in Gnat XR 977 with Brian Hoskins for my initial checkout. Thirty minutes of hearty aerobatics over the field with a few practice forced landings thrown in saw me officially on the team!

All we had to do now was get it right. Sprawled on our stomachs next to a Gnat; Bernie Scott nonchalantly advised me what to look for in my different formations, further down the line Malcolm and Neil were doing the same.

In Diamond, we line up the port nav light with the ejector seat symbol with the elevator leading edge in view – and the pilot's head looked about right.

For Fred – easier to say than Feathered Arrow – a whiff of power to line up the leading edge and see the second line of rivets on the fin. For Concorde – I was the port engine – getting startlingly close looked about right, the engine noise was incredibly loud and he left it to me to imagine what the whiplash would be like in turbulence.

For further finesse, Bernie pointed to a few choice bolts, rivets and hinges to line up. This was particularly useful if a bit of fudging was required – fudging is the vaguely imprecise art of deliberately flying out of position when the Leader piles on g to line up with crowd-front. It's where the visual aspect alters and if you flew perfectly in position, it looked wrong – so you flew subtly wrong to make it look right. With our new vocabulary and a host of things to line up on we were good to go.

All we had to do now was fly accurately in tight formation looping, rolling and bending around the field. After that, providing it went well and we hadn't disgraced ourselves and the rest of the team didn't regret choosing us, we would learn all the formation moves and get to use smoke. White smoke is produced by injecting diesel into the jet efflux and dye is added to give the distinctive red and blue colours. 'Smoke on go' means pressing the white smoke button. 'Colour on go' means you have to press the colour button... nothing can go wrong there!

Brian's flying was exceptionally smooth and he set a fast pace. My logbook shows us down to 500 feet after five days of flying three or four times a day.

Flying in tight formation requires bags of concentration, this induces tiredness and the more tired you are the easier it is to grey out under g and if you can't see it doesn't help matters. Despite wearing turning trousers to prevent blood pooling downwards it's essential to keep up anti-g tensing of the stomach to help keep the blood flowing to the brain and eyeball – which leaves you feeling even more whacked.

Wrung out, we watched the video replay drinking coffee grateful for a break and sugar rush before launching off again. After a day of this, it's obvious a sense of humour is vital particularly if it all goes to worms, Neil and Malcolm were terrific training partners; their upbeat sense of fun helped enormously. The Gnat might be nimble and a delight to fly but it's easy to over-control and provides entertainment for yourself especially if you screw up using airbrakes. Using them involves taking your hand off the throttle and grabbing the gear lever to partially lower the gear and if you had just taken off a handful of power – it got left off – so hands flash back to the throttle – and then you forgot about the airbrake...

The Leader uses voice inflexions to help us anticipate whatever's coming our way. A long 'Rooowllling out' means a slower rollout of a barrel roll compared to a faster one – and a high-pitched 'tightening' call means a rocketing heartbeat. The brain is remarkable in this respect and as formations become wider, audio cues allowed springloaded reactions in some positions or a fraction of a second delay in others.

The only time it didn't work was when Boss got whooping cough and we flew octagonal loops, only after a desperate team whip-round for a bottle of cough mixture did we fix the problem.

Outer positions 4 and 5 are considered dynamic. Malcolm had gone straight into the 4 slot which was a big ask and was coping well with the action. At the time we were living temporarily in South Cerney mess and in the evenings we were so tired we could barely keep our eyes open over dinner – but this was just the beginning – the real work was about to start.

Every pilot in a new role puts himself on a scale compared to the experienced hands and feels chuffed moving up the scale. But it doesn't work on the Reds, the scale's too large. Experienced hands were exceptional pilots with enough spare capacity to peel a banana while I thrashed around like a red yo-yo. But gradually you learn to snap-shot your position on the guy next to you, ignore others between you and the leader and take your cues from the lead's movements. You learn to subtly slow down by using g without touching power and revel in the coordination. We converge towards the same goal of creating the team of 1979 and although the standards might seem Olympian, respect grows and the magic of repetition under Brian's steady leading begins to pay off. Boss has been discussing options with the deputy lead Tim Curly an outstanding pilot, he is particularly well placed to do this having flown synchro himself, so between them, they have enormous experience.

Some minor precautions are taken: The sheer joy of a join-up loop takes some beating, it's used when the formation splits from a break and is all over the place. The leader runs in from the crowd rear smoking white for a loop and we hurl ourselves sideways at him to join up – it's awesome fun and I was always trying to outdo Neil Wharton on the opposite side who was incredibly quick. First attempts can be spirited shambles so the stem – Red 6 and 7 – are left out and only one side at a time practice it. This avoids the life-changing experience of hitting someone if you overcook it and fly through

Sporting manoeuvres are practised with height in hand like rollbacks in Big Seven with gear down. On command: the two aircraft on either side of the lead pull up sharply powering into a tight barrel roll outwards aiming to arc down onto the outside man who moves in as the next two pull up. It's lively and getting the roll rate correct is essential: If you roll too fast you lose sight of the man you are arcing down on and your heartbeat rockets, too slow and it's sloppy. Tragically it was flying roll-backs that Steve Noble crashed about the time of our selection so we practised up at 500 feet to get it right and then down nearer to 100 feet. For formation Twinkle Rolls we moved out half a slot. Our Gnats had a phenomenal rate of roll at something like 420 degrees a second but looked untidy on a rollout, so it was dropped.

Thanks to Brian's leading on November 23rd 1978 some five weeks after joining the Reds, we flew our first nine-ship.

About this time a circus guy is nominated to fly with you and an aircraft is allocated with both our names painted on it. Junior Technician George Scullion from the South of England is to fly with me – or as he put it – I would fly *his* jet when he wasn't fixing it. It was the start of a solid relationship with many humorous moments.

The Circus guys are sharp, George proved it to me one cold morning.

Boss led the front five off turning left shortly after take-off allowing Richard to lead the rear four-ship called 'Gypo' to join up – named after a swarthy complexioned chap who led it in the past, the call sign retained because of its distinctive and unambiguous nature.

As Red 9 in Gypo at 500 feet on the inside, I dragged power off on the join-up. Nicely in I shoved the power back up – nothing. Caramba! Promptly thumbing the engine relight button I rolled on more bank easing down gently. At pressing moments like this, it's a *must* to exit the formation. In a private aero sequence of my own, I managed to miss everyone in looking for the runway to land downwind. But it didn't look good. Pretty damned hopeless in fact so I told Lead I was ejecting.

For being polite in extremis I received an elegant 'Roger' while Boss carried on the practice. Choosing a field on the approach and whopping on some trim to crash in it, I reached for the handle. But amazingly the engine relit: I felt a good kick up the backside, got thirty-plus knots before I could blink and it quit, too fast now I crossed the threshold with 90 degrees of bank kicked it straight with a boot of rudder and smacked it on.

Streaming the brake chute and standing on the brakes I managed to turn off early to clear. Feeling chuffed, I leapt out to be greeted with the sight of a land rover towing a Gnat far too fast towards me. It was swaying from side to side alarmingly and behind it was another land rover pulling a Palouse starter. George leapt out asking what I'd done to his jet but never one to turn a gift horse down, I leapt in and started up. Within minutes I was airborne re-joining in a snappy loop feeling pleased to make the second half.

Last to land, I strolled in to sign in for my broken jet and the one George gave me. To be honest, I was hoping for a pat on the back for saving one – I'd already forgotten my poor throttle handling probably caused the problem in the first place: but laid across the tech logs was a white stick with 'blind bastard of the week' written on it. I signed in – puzzled –and ignoring George's mutterings about wrecking stuff I strolled mystified down to the debriefing room where the Boss was pointedly looking at his watch. The video rolled with everyone facing the front ignoring me – revealing a complete screw-up on my part.

The cameraman always switches to a jet in trouble and follows it.

There I was, a *smoking Red*. Guilty for not switching off the stuff when screwing up; when I hit the relight button I forgot to switch off the smoke. Mortified, I tried to ignore the hoots of laughter following my disgraceful approach and lousy landing. And instead of a pat on the back – I got a rollicking for poor smoke control and invited to eject next time and not play the hero!

The fact was our Gnats were timeworn and a complicated jet getting old is not a good combination. An instructor at RAF Valley flamed out on approach and ejected, and Little Ern, on his way to a meeting was late arriving because he ejected when his fuel proportioner failed and cut off his fuel. My modest contribution to the mounting carnage was to air test an aircraft fresh from a major overhaul and write it off. Ern was in the back and I managed to get into a pilot-induced oscillation after a firm touchdown, my jet seemed to have a mind of its own – I couldn't believe the nose-up attitude. Low on speed I whacked down hard and put part of the nose oleo up through the floor. It looked quite bent when we clambered out but Brian was remarkably calm when I told him I was sorry about wrecking one of his jets; although he brightened up when I said I could guarantee it was all my fault so there shouldn't be any problems with the others

About this time, I noticed a Stampe take-off as we landed to fly aerobatic over the field. They reminded me of Tiger Moth sessions years ago and intrigued I drove over to meet the pilot. A boyish-faced guy introduced himself as Vic Norman who said he was practising for a crack at the British Aero Championship. Invited back to the crew room he joined us for a coffee which started a long relationship between a talented hugely likeable guy with a burning ambition, and the Red Arrows. Vic was to prove highly successful and very much part of life at Kemble and to this day with the current Reds.

Venues for display are decided well in advance and when the season starts, everything is in place. One of the pilots is the lead Nav' for the display year with complete responsibility for planning and final coordination of timing and show clearance on the day. Stumpy was our lead nav' and Bernie is his deputy. Together they have made up a file for every display and they pitch up early to get the weather and sort things out. It's a vital role. Ray as team manager – callsign Mange – has been involved in display planning from the moment he arrived. As well as practising his commentary he has sorted hotel bookings, visits and public relations. Mange is the vital link between the team in the air and the public. Connected by radio, if he spots anything unusual or a threat to our safety he will inform the Boss. Normally he will fly in the spare jet or be driven or helicoptered in with our cameraman. It's a sought-after job but comes with a host of issues requiring attention and Ray was proving he was well up to it.

Weather usually decides the type of display the leader will fly: If the cloud base is high enough or patchy enough to fly a full display and contain a 2500-foot loop he will confirm 'fully'. If the weather prevents this but allows 1500 feet barrel rolls he will go for a 'rolling display'. But a cloud base of 500 feet with visibility just good enough to display can bring problems. In this case, we fly a Diamond Nine ILS approach and break out of cloud to fly a flat display keeping it low. As the season approaches, I notice we are not so tired. But sadly, our brothers on the international teams were feeling the pressure judging by the number of wreaths we sent out expressing our sadness for them losing comrades in practice.

When I asked who gave our public clearance I wasn't sure whether I was being razzed with the reply! – 'The Queen Mother our Commander in Chief, lunches with the AOC at RAF Leeming and after a few gins they watch us. If we get a thumbs up, the AOC says yes, we get to wear red suits and get a Christmas card every year.' She gave a thumbs up on 5th April 1979 and I felt immensely pleased that we new guys had made it thanks to a good Boss and remarkable team-mates.

And I get to wear a red suit with high hopes for an uneventful season...

And I get to wear a red suit with high hopes for an uneventful season...

chapter Forty-Four

First Season Hitches and Highs

Yeovilton 1979

JET XP 514 is in immaculate condition thanks to George Scullion who is flying Circus with me en route to Leck and talking non-stop about his soccer club Chelsea. From Leck, we transit to Aalburg for the team of 79's first international display and for the occasion George has given our names on the side an extra burnish (pointedly asking me not to hit any birds and damage his precious jet)

The weather over Holland is appalling as forecast with unbroken low clouds and passing rain showers. Just the sort of day to be on the ground sipping something warming: which is exactly what the French and Italian teams and everyone else thought since they hadn't even bothered to take the covers off their aeroplanes. For us, the situation was different halfway down a speedy Aalburg ILS doing 300 knots. The cloud base hung around 500 and the visibility underneath was reasonable so here we were. I was hanging grimly in Diamond Nine wishing the rain would go away when I heard Tim remark to the boss we were half a dot left on the localiser. Stum*py c*onfirmed and Boss commented he was laying off for the surface wind. How the hell the boys managed to fly in tight formation steady as a rock, read the instruments and have an off-the-cuff conversation fascinated me to the point of steadying me up. I even forgot about the rain.

Nine bright headlights appearing out of a ragged overcast was a welcome sight for a soaked crowd who hadn't expected anyone in their right minds to fly, never mind put on a display. Brian led superbly and deserved the salute of beer jugs from a grinning bunch of French and Italian team pilots braving the rain. He had not only flown smoothly in marginal weather but he had set the bar for the other world-class international teams.

The pace escalates, to keep a fine edge it's usual to display at weekends and practice a few times in the week. The experience hands set a high standard and nobody needs spurring on; the unstated rule is to be the best you can ever be. Tim Curly and the veterans of many displays show a hugely relaxed style and it's humbling the way they calmly shrug off a close encounter that's left you with a weak grin and a galloping pulse. I hope some of it rubs off.

The display routine begins with the team manager giving out arrangements for hotels, any engagements and pick-up times. It's all planned but I easily forget details and am happier being herded around like a lost ewe.

Stumpy and Bernie are the hardest workers sorting all display clearances and navigation, particularly on the morning of the first show or transit from Kemble. I just pitch up with a tennis bag and suit holder. These are piled into a bright red mini-van driven away by an energetic ground crew while I have a coffee and make a nuisance of myself asking Stumpy where we're going. He shoves a half-million map at me and I dig mine out and draw a line on it. The rest pile in with all the banter and darting wit of people with absolutely nothing to do and in quick time we have weather, frequencies and a line on a map to follow if we lose the Boss.

Usually, there is time for another coffee while the navs toil away at briefing details – this is always on time and sets the tone. Any rare almost genuine offers to help by new guys are largely ignored, because, in truth, everything works like a well-oiled Swiss watch without us.

The unruffled chat between Richard and the leader agreeing on the wind on the run-in and during a display is inspiring. I am used to it now but still smile behind my mask. Richard, synchro lead is painstakingly accurate with wind adjustments and he and Steve pithily converse to get their crosses smack in front of the crowd. Not generally known is a few displays are chosen deliberately in the middle of nowhere or at a difficult site and are 'in-season-practices'. These can be orphanages or charitable fetes and difficult for synchro and the leader to find never mind put on a full display. It keeps everyone sharp and is great for the kids. As the tempo increases the team spirit becomes stronger and the Circus guys star with repairs to our ageing jets.

George earns praise from Engineering when he made a temporary repair so good it could have been done in a hanger with proper equipment instead of in the pouring rain at Blackpool. I knew it was raining because he told me several times he got soaked during the night repairing the damage I caused – A big bumbling black crow was in my way running in to break after a display at Lakeside and I hit it with a terrific bang. Judging by the loud whistling noises coming from bent metal around the wretched thing's final resting place, I knew George wouldn't be happy – so after usual pleasantries and lame apologies, I departed for the hotel bar and left him to it. Undeterred by the carnage upfront: George cut up a large oil can and hammered it to the exact shape, pop riveted it on, sprayed it red and painstakingly polished it. It looked like a factory job.

Neil displayed great airmanship not to mention inspired flying when he suffered a complete hydraulic failure in a loop displaying at Manchester. It's a major emergency and he not only cleared the formation accurately but landed OK in difficult circumstances. Bernie got a brake failure ending up in the grass but despite everything the display has been judged a success by the crowds. Full credit has to go to Ray for commentating at times like this, turning eye-catching moments of drama into something positive takes a highly inventive style and a lot of brass neck.

Displays at Douglas Isle of Man seafront were hugely popular with Bikers. We met a few of the champion racers who recognised our enthusiasm for what we did as much as we respected their dedication to be the best. Barry Sheene flew with us before the season ended; Stumpy invited him to Kemble and flew him and we were delighted with the enthusiasm he showed in us. Barry was a colourful character interested in everything about the team and for a 750 cc and 500 cc Champion to express a vivid 'Wow!' flying with us was quite something. I asked him if he had as much trouble as I did hitting birds? He replied he totalled the front of his bike in Japan when a kamikaze pheasant hit him – just goes to show even famous people can be unfortunate too.

Summer was full-on and we new guys began to feel on top of the game, happy working to tight limits. Our fuel policy is three minutes endurance for the lowest man following a diversion from a full display. Steve number 7, uses the most fuel. After displaying at Wildenwrath the runway got blocked so Brian diverted to Koln Bonn. It was accurate planning by Stumpy with Steve smack on the fuel limit but we hadn't counted on a long taxi to the military side and I think only the lead made it to our parking place: most of us shut down on the taxiway with hardly any fuel.

Being short of fuel isn't new to any of us but if you have other pressing problems it can be awkward. Given that George thought I was a wrecking ball with wings, I still managed to sink to an all-time low on the amount of damage I caused on our way to Exeter for a display. A gear red warning light popped up passing Yeovilton Naval Station and a visual gear down inspection revealed the right gear remained up. Not good. Only one thing for it – bounce hard off the runway and shake it down. Simple.

Not wanting to block a display runway if things turned wormy I hung around the Yeovil circuit with the boys continuing to Exeter. I explained my problem to Yeovilton and could I have an ambulance please for my backseater because I was ejecting him if I couldn't get my gear down? Normal stuff to these boys. They were brilliant. And when I asked if they could put a helicopter or Hercules on their 'to-do list' so I could make the display at Exeter. No problem. They were on it.

Emergency gear checklists take no time with a show to make and I thumped it on so hard the Navy would have been proud of me. It was so brutal I almost dropped a wing but even after a teeth-shaker like that, we still had a red light on so I wasn't optimistic. George felt much the same about his jet being smashed into the ground but I tried a few more thump and go's until I was low on fuel: arcing steeply up to 1000 feet I told George to eject and not bother with an after-flight inspection.

He declined the offer. Said he was much happier staying with me. I wasn't surprised – I could tell he was a bit nervous the way he kept asking questions while I was toiling away. How many people were in China? On my estimate of a billion, how come they couldn't raise a team to beat Chelsea? (a sure sign of nerves – Chelsea was *never* that good): but with not enough fuel to argue or take a vote I called 'finals committed to land with less than 30 pounds fuel remaining.' This is about 25 seconds at circuit rate so we weren't exactly flush with the stuff.

Amazingly it worked out simply because the Gnat has outside ailerons. Keeping a boot load of rudder on I let the wing drop itself at slow speed keeping the stick hard over so the aileron took the brunt of all the scraping on the runway. We ended up ground looping on the grass with George completely dazed, so I nipped up behind him, made his seat safe and hoiked him out.

The doctor ran a stethoscope over him and said he was just a bit shocked – I told her it wasn't a problem, he was always like this whenever I bent his jet. Luckily, the Yeovilton hands had starred, our Hercules pitched up and bundling George inside we set off for Exeter.

The Herc' boys are fast at taxiing and they parked close to our Gnats – I hit the ground and sprinted to the Boss holding my thumb up, he gave a grin and the shortest brief ever – 'Wyn - Fully!'

I was tempted to ask why the briefs were so long but thought better of it. And I must say a cold beer never tasted so good after a nice tight display. George, bless him, after a few stiff whiskies to get over our jet being a mess was as right as rain apart from nightmares about me running him out of fuel.

With the end of the display season looming the new pilots joined for a look-see at the nine and eight slots. Tim Watts, a tall, fair-haired hugely likeable guy with a completely unflappable nature came from bona-jets. And Byron, an outgoing fellow Welshman with a wicked sense of humour knew everything about the Hawk having instructed on them at Valley. Two great additions to the team: Tim decided the job was definitely for him when the final downwards break put me in a perfect line over a famous golf course near Blackpool. Couldn't resist it! Four golfers putting on an elevated green is about as seductive as it gets for a wire-up level with the flag.

Our last display was fittingly at RAF Valley and our tribute to a great aircraft. It was a family day show and both my parents made it up to watch which pleased me no end. Brian kept us tight to the crowd and we signed off with a run-in at around 550 knots as low as we could, smoking and pulling up vertically to chase each other up a 'smoking chute'. A memorable ending to a season of 116 displays.

Last to land with eight on the runway I braked hard. It was a long taxi in and the brakes snatched a wee bit they were so hot. They can be susceptible to cracking with water leaking over them, but no puddles, no worries. Turning in I spotted Shep my father's sheepdog first – he never went anywhere without him – flicking my landing light on to say hello to my parents I shut down.

George grinned at my thumbs up, all good nothing to report. Superb. We had been a good pairing so it was appropriate to end on a high on our last display – fully serviceable!

Ambling over to greet my parents; Shep shot past me to investigate my red tractor – he had never seen an aircraft before – and the first thing he did was cock his leg over my red hot brakes, I heard a sharp pinging noise …

Long live the Gnat, a truly wonderful aircraft to fly but it's time to bring on the Hawk

photo Claire Hartley 'Long live the Gnat, a truly great aircraft to fly but its time to bring on the Hawk'

chapter Forty-Five

The Hawk

photo Claire Hartley

IAN MACKENZIE my astute basic jet instructor gave me excellent advice on never flying the Mark 1 of anything. He was right, but it was only the throttle response that wasn't up to scratch on the Hawk T mark 1.

The season is over and if asked to describe just one strong sentiment after the first year with the Reds, I would say I was staggered by the way our aeroplanes in their distinctive red colours captured public feelings. But officially it is now an aeroplane of the past. Our red Gnats might be a legend but the hard fact was we had six months to replace them with brand new untried Hawks keeping alive the public's expectations and their belief in us as their national team.

And it all fell on Brian Hoskin's shoulders as our leader.

Much easier for us guys in the formation engine room, all we had to do was learn to fly one and enjoy ourselves as we had on the Gnat. Fortunately, the Hawk is un-complicated with simple systems and easy checklists if things go wrong, which wasn't very often. Strapping into my first Hawk in October of '79 I reflected how astonishingly high the morale was on our team; the humour-charged atmosphere in the crew room was infectious to everyone. Almost!

Wing Commander Holliday the chief instructor sitting in the back was a friendly soul with a precise air about him. His exactness made me realise with a start that I was a student again under normal rules which was a bit of a shock. I had no difficulties with this except I had forgotten what normal flying was …

The Fleet Air Arm had been a thrusting sort of life. Holding your own in lousy weather on non-diversion ops with first-rate hard-assed flyers who didn't bat an eyelid hooking onto a slippery deck without much fuel: wasn't exactly normal. Leaving directly to join the Reds, who were amazingly tight-lipped about doing things solo that you wouldn't normally get away with (court-martialled) probably wasn't the best preparation to be a model student either. I tried hard to be one for two hours and forty minutes of dual instruction. But in the last five minutes, I didn't try at all. We had meticulously covered stalls, spins and 'edgy' areas and I thought the Hawk was well designed and easy to fly. And If I could fly it without breaking it – anyone could. But I was frustrated with prim circuits at exact speeds and precise power settings. I needed a more robust bash because we were shortly going to fly it to the limit, so I asked if I could have a slightly more operational crack at circuits?

'Certainly old chap.'

'Valley Red 3 is 30 seconds and confirm the circuit is clear.'

'Clear.'

The Hawk was beautifully light on the controls, pulling up into the vertical from a slightly non-standard 50 feet circuit and pointing down for a simulated bomb burst was completely effortless. We had a winner here. With roll-backs in mind, I altered the normal circuit slightly by doing several sharp barrel rolls downwind to get the hang of rolling with the gear down. A deep silence from the rear – which I took for approval – encouraged an impromptu Derry turn onto finals. There were no hidden problems that I could see providing you felt the stall buffet through flap vibrations. Great jet! Holliday was silently sporting about me sounding out a few strictly forbidden areas and after a quick solo around the parish: it was time for something serious.

Messrs Benson and White instructed me with great enthusiasm but my ears pricked forward like a terrier when they discussed throttle handling. A definite lag in acceleration response – bad news for formation aerobatics.

The team had picked it up immediately and Hoss was on it fast. We gave it a go flying a nine ship which surprised the instructors as we had so few hours. Predictably formation keeping was rubbish in the vertical which left us feeling glum. However, BAE was now deep into a solution with the Adour engine boys; they knew the export potential was huge and our pride and credibility were at stake. Hoss remained unflappable throughout and once the engine response time was sorted: our formation keeping was instantly better and any doubts about the Hawk not being as good as the Gnat disappeared.

Squadrons frequently change aircraft without fuss and the Reds were no exception. Certainly, the dye team looked a lot cleaner! Reloading coloured dye into the Hawk was easier but on the other hand, the engine exhaust was cooler than the Gnat's so the dye didn't emulsify well and the colour wasn't as bright. Bob Lewis promptly sorted it and moved on to the next task.

Brian Hoskins, a superb leader in the last year of the Gnat excelled himself on the Hawk. We needed more sky to fit our display into and he and synchro sorted out minimum safe heights for loops and opposition moves; everybody had a say in changes to the formation and the Reds were back in business again.

To be conservative the cloud base needed to be approximately 4000 feet for a 'Fully', below that down to 2500 feet we flew a rolling display and below that a flat display down to a 'reasonable height'.

The sweepback of the leading edge was roughly half that of the Gnat with a wider wingspan so we needed new reference points for our visual trigonometry. Amazingly we got it right with only about one foot of a sorry-looking crumpled wingtip between us.

Nonchalantly pitching up to fly and doing little else ended when Bernie chose me as his deputy team navigator. Despite having to work again I was secretly pleased because very little was known about the Hawk in formation aerobatics to a world-class level. The US Navy was showing interest so we had to get moving. Especially as the new Alpha Jet made by Germany and France was on the scene with the French being the predominant partner. Les Bleus were determined to show us up.

One winter morning waiting for the fog to lift, Boss asked for thoughts on changing the team's name along with the aircraft. 'The Black Knights' with our jets in all black colours was a strong suggestion from a very senior level. It was entirely up to us and over a coffee and unusually strong banter the vote was for the Red Arrows to continue. Everyone on our team had aspired to join the Reds, the reputation of the team and our red aircraft was built on the efforts of distinguished flyers, some of whom had died building that reputation. The name Red Arrows was a tribute to them. It was going to stay.

Bernie and I started from scratch with display planning and although winter temperatures gave good performance it was sensible to look at options for a hot environment. Playing around with the fuel weight was the obvious solution until we had more experience; even starting a display with less fuel to keep the weight down we could still divert with 15 minutes of endurance which was sheer luxury. A training period of ten days in Cyprus for good weather continuity was the answer to the maiden's prayer for us. We were in a new league with the Hawk's fuel economy giving an easy range of 1000 nautical miles plus. This with our integral air starter meant display venues were almost worldwide as we needed less ground equipment and support.

The winter of 1980 was challenging. And not just the flying either. Ray as manager dealt with the media's gathering interest, Raymond Baxter flew with us filming for the BBC and my log-book shows Richard Cooke, the talented photographer I knew for his shots off *Ark Royal* flew with us. His creative style was a brilliant success and won many awards. It was clear we were in a new commercial dimension fronting a new aircraft for British industry with Ray bearing the brunt of upped public relations. It was time for officially trade-marking the Reds and luckily we had two stalwarts respected by the whole team giving us marvellous support: David Boyce from Lloyds of London for insurance and Eric Ward advising strongly on the commercial front.

An extraordinary few are involved in organising the display season. Navs – plus the team manager and team engineer are overseen by the Leader whose word is final. After a few days of brainstorming the list is passed to the MOD and the carpet to the ops board is worn thin with guys strolling in for updates. As lead navigator, Bernie is responsible for obtaining weather, planning transits, meeting show timings and preparing maps for the leader. He will be in touch with show officials chasing down details and sorting frequencies. Political and diplomatic clearances are our responsibility and initiated via the station switchboard manned by two great ladies who seem to know every Air Attache in the world! Dividing the shows between us we got cracking and I tried not to think of next year when I take over responsibility for navigation. All we needed was a spell of good weather in Cyprus to get it all together. Exercise Springhawk was borne…

After training in Cyprus, we were ready to stretch our wings in front of the public.

Return from Springhawk

chapter Forty-Six

Meeting Expectations

Opening Loop

EXACT POSITION inverted over Sywell

On April 6th '1980, Red Leader Squadron Leader Brian Hoskins calls 'smoke on go.' Sweeping in from crowd-rear streaming white smoke he pulls up in Diamond Nine for our arrival loop and our first UK public display is underway. Over the top, Ray radios our new Hawks have drawn spontaneous cheers and applause from the crowd. It's a big moment – the public had accepted the Hawk.

I was always proud of the Reds – they're a public statement to the taxpayer of excellence achieved by the RAF which I joined as a mere boy. The standard and ethos portrayed by engineers and support staff are as good as anyone can achieve in the world and Trade Attaches claimed the Reds provided proof positive of the precision of British products is reflected in sales. In other words, we were worth having around. A new experience for me. We got cracking.

Competition to join the team was fierce as ever: John Myers and Ron Trinder from Buccaneers came along. I flew with Ron on a solo day and couldn't help thinking the RAF had immense talent in-depth with guys like these. Steve Riley ex- Lightning and Phantoms, Pete John on Buccs both ex-Ark and Ian Huzzard a Hunter chum not to forget Neil Matheson who flew a mean Lightning display could fill any formation. Other hands probably felt the same and all I could do was wish every single one of them good luck!

The Circus liked the efficient Hawk and looked forward to cleaner overalls. Bernie and I had the performance data sorted out in Cyprus, Ray whipped his commentary into shape and sounded like a Hawk salesman for BAE, and wherever we stayed, we used a room to view the video and critique before we dined, no matter how knackered we were. 1980 was a blur. Bernie and I cobbled together performance data, the PR side warmed up, and we flew in formation with Concorde for her first Hawk formation shots by Arthur Gibson. After several early displays, we transited to Biggin Hill on the 17th of May '80 to display at Brighton and Biggin air show. The weather was good with no turbulence and the display over Brighton was going well. Exceptionally well, right up until a dim yachtsman decided to sail his boat into a cleared space on the show line. Steve was smack on this line at 35 feet and reacted fast when his wing clipped the top of the mast. Ejecting almost inverted, his aircraft XX 262 crashed in the sea thankfully missing the Palace Pier and spectators.

Ray called that Steve had ejected and we all held our breath knowing what it was like down there with boats and people everywhere. Ray quickly confirmed Steve was OK – no one hurt and Boss staunchly decided to carry on the display. Being Red Three next to him I remember moving in tight and a glance around revealed everyone was doing the same – a support thing!

Landing at Biggin we prepared for an eight-ship display. The media heard of the crash but were good. Every year we were invited to ATV News at Ten and met senior media directors for drinks. One senior director was present and he did us the courtesy of leaving us alone while we sorted out a few moves. Preparation for the season includes reviewing a forced eight ship and prime movers. Change of call-sign and revised positions are noted on a nicely typed sheet of paper, then stuffed into a pocket and forgotten about because it can never happen to you! After welcoming a dripping wet Steve back and helping sort out his room and hospital check, I sat on the wing of my jet gazing at a tattered piece of paper trying to work out what the grubby arrows meant. Amazingly it worked, Brian led with enormous panache in a difficult situation, the public loved it and best of all, Steve's back was in good shape a few weeks later.

The next two years would be crucial in displaying the Hawk and giving a patriotic shout for British manufacturing. Farnborough International Air Show would showcase us to the aviation world as would Greenham Common and Fairford, both International Air Tattoos. On these venues, the Alpha jet moved into direct competition with us as a trainer. Aware the contracts were enormous, we were perfectly happy to take on any national team re-equipped with the Alpha in terms of display ethos. The accepted rule amongst world-class teams is that the national team will either open or preferably close an international show in their country. It's more prestigious to close and leave a lasting patriotic impression on the nation. We never displayed at the Paris Air Show and the French, ever patriotic would never expect it.

In 1980 we inched improvements to our displays in every possible way. Displaying from my home turf at Llanidloes across to Geneva, there was a discernible feeling that the excellence of the Hawk was apparent to the world at large. It was sturdy, dependable and reliable (I inadvertently pulled 12 g with no problems mixing it with two A10s on a solo trip), we performed for the Queen Mother in the summer and in early September, Farnborough International Air Show days focused our attention. It was here that we had to put on the display of our lives in the hallowed airspace that for generations had seen new aircraft perform. No pressure!

For Malcolm, Nei and myself it was an ambition fulfilled. Boss led brilliantly to within an inch of the crowd line and every team pilot flew with that almost unnerving exactness you find when you are proud of what you are doing. I swear Malcolm in the Red 2 slot opposite me never moved out of position one inch!

Venues whizzed by. Our last in Europe was Auf Den Dumpel and Hiltzingen where the site is on top of a hill. It's wonderful for wiring down the valleys and flying inverted through someone's orchard. I almost got an apple strike on that one! After enormous steins of German brew overnight at Frierdrichsafen, we finished at Cosford and the 1980 season was over with 119 official displays under our belts.

It was time to consolidate our positions. Synchro to remain as a pair retaining expertise and the Brass wanted a return to three new pilots a year. This meant an extended tour for one and a shortening for another. Unfortunately, Malcolm drew the short straw and Neil and I batted on. It was a wrench for us all to see Malcolm go early but he went on to instruct superbly at weapons school and they gained an immaculate flyer.

John Myers joined from Buccs and Ian Huzzard, an old chum from Hunter days, arrived with Henry De Coursier from Harriers to form the team of 1981 along with a new Engineer Graham Nisbet.

I was sorry to see Richard Thomas and Bernie leave. Those who chose you for the team always have a place in your psyche. Richard phoning me up was a magic memory and flying next to Bernie was always fun. He passed on a wagon load of information on planning so as soon as the Boss approved me as the lead navigator I started musing over a Middle East tour. Rumours were strong, a framework had to be in position. My slot was Red 3 again next to the Leader: it made the nav easier and also meant a calmer training period with more time to relax. Exercising a friend's hunters at Minchinhampton Common was as good as flying the Hawk, they were forward-going and fun to ride out in Cirencester Park with my wife Jane. Viewing what was coming I needed that.

XX 253, George Scullion's jet and mine performed well with no failures and met his exacting standards with only a few more or less sparrow-sized strikes through the season. I swopped it for XX 266 for the 1981 season with dire warnings not to bend it!

chapter Forty-Seven

A Royal Welcome

Salaam Formation

OUR NEW display filing system worked well. Under my left buttock, the height of the files equalled those under my right one. I had decided this intimate balance was crucial in keeping my back straight if I had to eject in a welter of files and bits of paper. Duplicates were with Uncle George in the office so Byron Walters the new deputy navigator would just have to sort it out if the worst came to the worst. Our little nav team with its easy camaraderie now had an Operations Corporal and between us, we sorted out the navigation of the entire tour and coming display season.

Winter training for 81 had been full-on for John, Ian and Henry our new guys to get up to scratch a month early. The plan was to lob over to Cyprus for our annual Springhawk training session; fly our backsides off there for two weeks, get everything together and then springboard across Saudi for a tour of the Emirates, and Bahrain and then return across Saudi to Jordan. All before our normal season!

Boss just let me get on with it and appreciating his trust I grabbed a large two-million scale chart and started drawing tracks everywhere. With enviable access to Bill Bedford Chief Test pilot BAE and liaison with Air and Trade Attachés from various countries, a plan emerged. It would be modified by events in Cyprus but a framework was in place to dovetail with International Air Displays in Abu Dhabi, Dubai, Sharjah and Bahrain. The Royal Jordanian Air Force would host us in Jordan and Ray as our manager took care of all arrangements with BAE and MOD on top of his usual pre-season display frenzy. A big ask.

BAE Sales Team, wise in the way of Gallic selling made discrete inquiries about the Alpha Jet competing against us in the Middle East. Exporting arms and aeroplanes forms part of France's foreign policy and to help their case along they printed hundreds of Hawk trade pamphlets and sales data identical to ours which they thoughtfully distributed to governments everywhere. Complete with wrong performance figures and some Gallic bending of facts they amazingly put the AlphaJet way out front! Selling aeroplanes was a whole new ball game to me and I must say the French claiming their subsonic jet represented the best qualities for supersonic training was pretty impressive salesmanship. Their wiliness to trump our virtuous Brit approach by telling porkies was going to be fun to watch.

On March 2nd '81, all embassy phone numbers, diplomatic reference numbers, political clearance notes and site maps were filed under my left buttock. They balanced neatly with various charts to get us to Cyprus and a treasured two million scale map of the big picture under the other one.

A support Hercules departing Brize had to be tasked early bringing our plans forward. I was at my Mother in Law's funeral in Yorkshire at the time and had to leave my wife there. Being ex WRAF she was marvellous: and when Brian twirled his finger for a formation wind up and called 'Rolling' I was determined everything would go well. And as wheels clunked up, I changed to VHF frequency for ATC to clear us on our first leg to Pisa, Italy. A quick refuel and on to Tanagra Hellenic Air Force base for the night. Springhawk was underway!

Training with guaranteed weather gets any wrinkles sorted out and gives us a chance to get some zip into a display. On 3rd March Boss started formation practice and by 19th March we were ready for our departure on tour with Public Display Clearance displaying at Limassol, Akrotiri and Dhekelia with two days of photoshoots. A busy Springhawk – the RAF gets its money's worth and if anyone thinks it's just poncing around in a red suit don't bother applying! The new guys were brilliant taking everything in their stride and Byron and I went into a huddle. Region's politics required flying airways under Damascus control south to Jordan for Tabuk in Saudi for refuel, then cross to Muharraq for a next day transit to Abu Dhabi for display start. Small problem, TACAN was not an airway aid and we had no trucky navigational kit. It's a meticulous process using en-route charts tracking airways and transposing routes onto a topographic chart for the Boss to map read. Everything was dead reckoning twirling Dalton computer picking off drift angles, ground speed and all-important Mach number – lead's aircraft was promised a VOR sometime. The accepted rule is to leave the navs alone and whip up a standard NATO coffee every hour; the importance of the trip became evident when the Air Officer Training quietly pitched up next to me beavering away. I muttered to Byron we could do with a 35,000 wind and temperature at the Saudi border and minutes later a gentle cough came from the AVM as he slid a 'Flash' message across to me with winds and temperatures. God Bless good senior officers, without a word spoken I knew any support we wanted was right there.

March 20, Akrotiri. Wheels up, sliding away from the leader for wriggle room, I called 'Three to Damascus,' switched to VHF and they cleared us all the way – nav remains on VHF listens out on UHF and makes all calls. Despite all the planning it's still good to get a thumbs up from the leader as you shadow him and know the sums are working. Things went well up until checking with Tabuk Approach. We didn't need to check in to see the problem.

This wasn't a full-blooded Haboob dust storm I had seen years ago but a sandstorm can look innocuous viewed vertically but horizontal visibility is much-reduced, and blowing sand gives problems. We couldn't start. When the Circus boys inspected the auxiliary power unit they found the sand had clogged various bits. Normally not a problem with an overnight stay and Circus working flat out to get us away first thing but not the case..

Tabuk, situated in the North-Eastern military district is hugely strategic and the most isolated and largest Saudi airbase. This makes it politically important. And if Maggie Thatcher hadn't upset the Saudis with some remark recently it would have been fine but she had, so we were unwelcome and had to leave now! Almost comical as we couldn't start. The realities of our technical situation handled in this manner began to seep through to the Saudi Command as being un-airman and a reflection on them – but it wasn't the British way to make a fuss and we managed some shuteye while the Circus guys laboured on.

A day behind schedule we needed to get some traction on events so it was pre-dawn hoist of harness and in the early hours of March 21st at the first crack of light we took off into a rising sun for Bahrain. Crossing the An Nafud desert was dramatic but as we neared Muharraq sea fog rolled in earlier than forecast so Boss diverted to Abu Dhabi where we were going to display anyway. I still shudder over the fuel calculations but Neil I think was the low man smack on minimums gliding in over the desert hardly daring to touch the throttle.

No rest. A bloody long day. We displayed at Abu Dhabi, with Boss winding it up beautifully keeping it tight, landed had a cuppa, met local royal dignitaries, launched for Dubai to display Sharja and landed back at Dubai feeling we had done our best.

I barely remember a post-flight drink viewing the videos. Ray was as tired as we were but he batted on stoically and deserves credit for not getting upset with the French... We heard a strong rumour a Sheik in favour of buying the Hawk couldn't attend because someone shot his house boy and the chap who pitched up instead favoured the Alpha – but as an ordinary Red unplagued by any sales pitch, I just went to sleep. We had barely stopped since rolling down Kemble runway on 2nd March. I couldn't even remember where I was or the name of the hotel. Next morning we transited to Muharraq, displayed Bahrain and Byron and I unrolled the grubby-looking masterplan onto my wing one more time…

Chas Montgomery Captain of the support Hercules strolled over to see how to help. Chas had been brilliant obtaining winds and routing clearances for us and frankly, he and the Hercules were now part of the team. Our transit back via Tabuk was markedly different; It was smiles all around now and we were soon ready to go. But it came as no surprise when two armed Lightnings scrambled in a show of dust to escort us to the border – bit silly doing that to RAF pilots. Two of them and ten of us. Boss kept straight on course and we loosened up as normal but this time by unspoken agreement kept miles apart. This forced our heavy friends to choose who to stick behind – trying to ignore Hawks sneaking into their six and gaily barrel-rolling around their ears

Amman's Queen Alia Airport welcomed us and the Royal Jordanian Air Force promptly took charge on the personal orders of King Hussein. Whisking us away for a bit of R & R they footed the bill and I've no idea where we went to sleep for the next few days. Back in the saddle after four days, we flew to King Hussein Air Base Mafraq in the north for display. It's an old established Academy and interestingly once a RAF armoured car base.

Jordanians were always wary of assassination attempts on his Majesty and on the 28th March after a to and fro of a Hussein look-alike in a helicopter decoy, we eventually displayed for him at Amman. Afterwards, we met the King. He had a formidable reputation as a pilot himself and being a long friend of the Reds was charm itself. We spent time with the Royal Family and flew the Royal Princes, each of us received an elegant watch inscribed with HM's Hashemite crown to mark the occasion and Boss got a medal, so our display must have been up to scratch.

Before the tour, there were discrete murmurings of our part in sales of the Hawk with a briefing on what to expect from the BAE sales team and Rolls Royce. They promised a strong discrete presence at any official reception we attended. I was busy planning my small part of the big picture at the time but

I recollect it envisaged us engaged in conversation by royalty or a political dignitary and then passed on to a military type. He would probably ask enthusiastic non-awkward questions about flying and shooting at someone or something which would naturally lead to the question of how much? A winning smile and a discrete gesture would bring immediate support from the top-end sales of BAE and RR. Theirs was a more sophisticated approach than a wild-ass guess by us: apparently, you don't sell someone a brand new squadron of highly polished Hawks and ask for a few million pounds cheque. Even sharper minds than the Gallic ones churning out tweaked data in France were at full throttle on this one.

It seems International selling was a refined business involving discrete options regarding the price of crude oil, importing sticky dates and camels and possibly oil drilling rights and trade offsets – I may have got some of that wrong because I was concerned about getting the fuel calculations right and not running out of the stuff crossing the mighty An Nafud desert at the time. Watching that level of diplomacy at work and the efforts put in by Rolls Royce and British Aerospace impressed me, and to be honest, quite humbling to think all I did was enjoy myself. With maybe a spot of planning thrown in.

But the Hawk was on Iraq's shopping list. They invaded Iran in September 1980 and wanted a trainer with huge expansion plans in mind for their Air Force. We met senior officers at a discrete base out in the desert. The Jordanian Officers were charmingly attentive, I asked about two long lines of F104 Starfighters obviously grounded with blanks and covers. He shrugged 'no spares' – what a waste. I got chatting to an Iraq Major who gave me a blow-by-blow account with nifty hand moves of his latest dog fight with an Iranian F4. I guessed he won! Needless to say, we had powerful sales backup and I sensed natural respect from the Iraqi senior officers some of whom flew in our Hawks. On March 31st after displaying for King Hussein we said goodbye to our generous friends in the Jordanian Air Force and left for Cyprus to give a farewell display for Akrotiri. It was time to go home. We went via Tanagra Hellenic Airforce base after displaying in

Athens and on April 2nd we touched down at Kemble exactly one month after we had departed. Phew!

chapter Forty-Eight

Farewells and a Rewarding Year

DURING THE opening stages of the Falklands conflict, I expressed a wish to leave the Arrows and fly Sea Harriers with the Navy if the war showed signs of being a long engagement. Boss, always a good listener promptly evicted me from his office to go and do something useful telling me I was already on a Navy list. Satisfied, I headed straight for the display folders to sort out planning our confirmed shows.

We kicked off the season by refreshing flat displays and roll-backs, before launching into the season of 81. Being Brian's last season as the leader, everyone wanted it to go well for him. His foresight in training the team up early anticipating a tour in the Middle East was paying off because our new guys Ian Huzzard, Henry de Coursier and John Myers were already experienced hands. With over 125 displays planned, there was even more going on below the surface than usual. New guys needed selecting and a new leader needed to be in position. The Reds choose their team leader from a shortlist. They can and have said, no, in the past to any strong suggestions from above they are not happy with – and that's final. The new leader will have Red's experience so much is known about him and it's usually a close-run thing but it's the team who decides who's in front.

Displays come first and this works well if time is needed to reflect on choices. The atmosphere is as focused and pragmatic as it gets. After flying a display in ghastly weather and rubbish turbulence, if an equally knackered, sweating team-mate asks if you would be happy following 'so and so' around the skies in that lot – it's not a question to take lightly. Spring of '81 would see a raft of changes because world interest in the Hawk as a new generation trainer was growing fast. The patriotic impulse to put on the best show possible with a British aircraft was strong and came from all sides with the media wishing us well. The US

Navy contract was important for national engineering prestige but it was also an important fiscal year because European Air Forces were re-equipping. Dassault-Dornier pushed the Alpha Jet hard as did the Italians with their Aermacchi 399 MB. It was important to get the message across at the big international shows like Greenham Common and Fairford. The Fairford organisers were a relaxed friendly bunch and organised their International Air Tatoo superbly. However, not everyone in the display world had the same constancy in support which led to something like a pantomime.

I was up to my neck planning when Boss asked me to fly down to Greenham Common to sort out an issue. The organisers insisted on speaking face to face for some reason. Sounded odd to me and Ray wisely advised I checked the files. Good job too. Something was afoot. Normally we displayed at BAE Manchester Woodford, landed Greenham, refuelled, closed the show then landed at home to position for further displays. Unusually, at their request, we were overnighting so it all hinged on closing the show each day – also a great chance to meet our opposite international numbers. A Patrouille de France pilot I knew well always looked me up to enjoy a glass of wine or two. The last time we swopped our flying suits before a display.

Greenham attracts national display teams from around the globe; these are great gate-pullers but on this occasion, it was a good chance for the public to see their national team flying a British contender that could bring welcome jobs.

Uncle George phoned my ETA to Greenham and grabbing the display list and diary I filed them in the usual place and flew off. On landing, I was instructed to taxi to a remote part of the field, presumably for security and wait for transport. After 35 minutes I fired up the radio and asked if there were any problems. The controller was apologetic, she had called the organisers, twice. I was to wait.

It was a definite snub and having a natural proclivity for poor judgement in situations like this I decided to have some fun. I filed a flight plan back to Kemble to get a practice in asking politely if they would patch the departure time through to the display office. It was thirty-five minutes hence plus five minutes for a sense of humour failure. A car appeared by magic, I was shown into a room cluttered with faces sporting a range of smirks but a couple tinged with embarrassment gave the game away. Without any introductions, I was informed the Arrows would be included in the show with the French, Italians and Swiss which I thought pretty decent of them – but they had decided the honour of closing *their* show was going to the French and Italians.

The atmosphere spoke for itself so pointless adding my voice to it. Trivialising the team with calculated slights is nothing new – it happens! But it's unfitting to respond in kind and emphatically not our way – far better to be amiable and helpful so I simply introduced myself as responsible for planning. Fishing out the display list sent to MOD previously agreed by them and heavily dented from my left buttock, I laid it firmly on the furthest table away politely informing them my word was final and suggested they compare times – and was there anything else I could help with? The scramble of feet towards the list as I closed the door said it all. Being a team navigator was getting to be fun!

it seemed unfitting to let the public down by Greenham effectively cancelling our display at Woodford just to allow a foreign team to close their show. The footnote was comical and predictable. Lots of trophies! Best smoke, best display etc. And probably due entirely to my diplomacy in off-siding a few people, each trophy was pointedly given to our international friends which we applauded with gusto. They saw through it immediately and one distinctly unsteady French pilot with a bottle, two glasses and a huge grin hailed me and told me being praised by the best team in the world deserved the best wine and produced an excellent year Margeaux ...

It was full-on from June with a memorable display in front of the Queen on the 17th of July opening the brand new Humber Bridge. In the same month, a new shape appeared in the skies over Kemble in the form of the Prince of Wales Feathers. Requested to create something for displays over Carmarthen Castle, Boss sorted out a vertical upwards break smoking white with three vics of three diverging outwards over the top. Being Welsh I was proud of that.

July and August after our early tour seemed delightfully never-ending with displays throughout Europe and UK. It was exquisitely demanding but this is where the magic of repetition and mutual trust creeps over any tiredness. Hands move over throttle and stick without any conscious thought, eyes snapshot position, movement and speed to correct automatically. Nobody would ever say it – we were far from philosophical: but there comes a time every year when we feel we no longer have an expensive heap of nuts and bolts strapped to our backsides in the form of a sleek red shape. Movement guided by the subtlest of touches makes for an easy feeling of attachment to the formation – and the Hawk didn't break as easy as a Gnat either!

Welcome visitors spiced up the year. The Shuttleworth Collection boys flew in for a day to Kemble with all manner of vintage aircraft from a bye-gone era, they impress everyone with their boyish enthusiasm of young men and the skill of veteran pilots. A great bunch! Noel Edmonds, one of the first disc jockeys on radio Luxemebrg and a popular celebrity with the BBC joined us for a few days with a film crew. He was a good hand, as sharp as they come and hilarious with it. Interested in everything we did he loved the casual banter and flying with us. A shame when he left.

The end of the 1981 season was on us with usual suddenness going from flat out to a final display at Andover. Always special for me after my early clumsy efforts in Jackaroos.

Changing the leader is a time for reflection. If you leave with him it's probably easier! Steve Johnson and Neil Wharton two solid outstanding natural pilots and great teammates did so. Brian who introduced the Hawk, mentored and led us for 125 displays this year, flew over 360 official shows in three years – not to mention his previous team time – handed over to John Blackwell. And as luck would have it Brian chose to extend my tour with the team so it was a time of mixed emotions. Couldn't wait to start the season again but on the other hand sad to see a unique Boss leave.

BAE with MacDonnell Douglas won the US navy contract to produce a carrier-capable Hawk. The T45A. It became the highly successful Goshawk and I have no doubt whatsoever that

Brian played a big part in this by showing what the Hawk could do. Not forgetting our team manager Ray who not only had to join the dots with the substantial increased PR associated with a new type but also kept up constant liaison with BAE.

John Blackwell took over with two new pilots Tim Miller and Phil Tolman joining him. I moved out to the Red 5 slot on request; it's a fun dynamic slot and flew opposite Ian Huzzard Red 4 enjoying a good pairing. We got to know each other's flying well and the timing of formation changes to the second where any slight delay triggers an immediate awareness of a problem. When Ian didn't acknowledge a movers-call we promptly looked across at each other, nodded, and continued with him executing changes on an enthusiastic head nod from me. I felt pleased we hacked it together but we were completely outshone a few days later by Tim Watts Red 6 synchro lead. After leaving his flying helmet behind in the hotel for reasons best known to himself, he grabbed the team spare strapped in and found the microphone unserviceable on check-in. He gave a bleep on the tone button to alert Henry de Courcier Red 7, gave him a thumb down and they did the entire display and synchro routine using bleeps! Great display of airmanship by the pair of them.

After busy preceding years it felt good concentrating on pure flying, training up new guys leaving most of the navigation to others. John Blackwell threw himself into the lead with enormous enthusiasm proving to be an instinctive leader and a good choice for the team. But it wasn't a question of 'mission complete' bringing in a new type of aircraft. New ground broken by the current team paved the way for what is now a form of sponsorship. Something unheard of in the RAF. In this, the Team were fortunate in knowing Eric Ward a Director of a large textile company who was introduced through a client whose brother was a Team pilot. Eric quickly spotted the pilots might be the epitome of excellence in the air but they dressed in a somewhat uninspiring fashion in a civilian 'uniform' of mediocre quality. This weakness, already noted by Ray as team manager was in addition to a shambles of poor-quality memorabilia being sold by anyone who decided to be an airshow trader. To illustrate the point Marks and Spencer were marketing Red Arrows T-Shirts without bothering to inform the Team.

To be fair, this was a completely new area of involvement by the Commandant CFS. He was suddenly being confronted with a dearth of commercial awareness due entirely to the worldwide success of the Team.

In Eric's opinion, legal protection was needed to prevent any serious damage to the image of a national team. Justifiably and quite independently Ray set out with Eric to trademark the national team. Now a trusted 'Friend of the Reds' Eric had the ear of Air Marshal Sir John Sutton C in C Support Command who very much approved their efforts as an essential step. Undoubtedly the level of sponsorship enjoyed by current teams is due entirely to the foresight and hard work of both Eric Ward and Ray Thilthorpe.

However, these visionary aspects needed focus, so on the Bosses instructions following the team's request – I ditched most of the planning to manage the pilots on the road to help out. Not a choice of mine being a lazy protagonist of 'you stick to your end and me to mine' but I just got on with it trying not to ruffle too many of the team manager's feathers because it made sense.

The early '82 training was unusually busy with special flypasts and displays to commemorate the victorious return of the Falklands War forces which was particularly gratifying. We flew impromptu displays to troops who had been wounded in battle and invited them to our home at Kemble. A pleasure to fly for them and a privilege to meet them. After one humdinger, flying lower than ever and socking it to them – which would have grounded us forever if the Air Board had witnessed it, I crouched beside a para who had lost limbs at Goose Green. He enjoyed a beer watching us. I detected the same indomitable cheerful spirit I remember from old days in Africa and the Middle East shining through and we had a bloody good laugh together. I just hoped we contributed something however small towards the thanks the nation felt for them. Brave guys.

As far as we experienced hands were concerned the whole season was going to be a success due to John's enthusiasm, humour and the readiness of the whole team to gel around our new leader. John put his stamp on the team and as ever, Springhawk in Cyprus gave us the extra edge of good weather to tidy up for the season ahead. It's where the character of the 1982 Team began to emerge. It's also a good time for getting to know new Circus Blues well and very much up to the pilots to break the ice. I normally crack a few jokes and give the impression of never causing any extra work by breaking aeroplanes – I normally get away with it unless Scully my first blue gets to them first – and one crisp morning under a clear blue Cyprus sky leaning on the wing of the jet I was retelling an old joke remodelled around Bucc times:

'ATC receive a call asking for a time check please'. Tower responds ''Who's calling? The crew replies What difference does it make? 'Tower replies 'It makes a lot of difference. If it's a British Airways flight it's 3 o'clock. If it's the RAF Ascot chaps in Hercs it is 1500 hrs. If it's a Navy aircraft, it's 6 bells, If it's an Army aircraft, the big hand is on the 12 and the little hand is on the 3. If it's a Buccaneer crew like you, it's Thursday afternoon and 120 minutes to 'Happy Hour'.

The flying joke went down reasonably well but I regretted my reckless remark of not damaging a jet for ages because we took off and I hit this huge eagle head-on just as John pulled into the vertical for a Diamond Loop. Biggest eagle I ever clapped eyes on and a hell of a bang as the top of the windscreen and canopy took it. The noise can be a bit off-putting and lots of big birdie blood streaks don't help the view much – or steady formation for that matter so best to gracefully leave the formation. If the engine still works everything is pretty much OK and the situations not worth complaining about. The only minor problem I had was the shockwave from impact blew off the whole detonation cord around the forward canopy which flopped nicely around my helmet!

The thing is designed to detonate and shatter the canopy on ejection so a pointless option now as I would just blow my head off. Careful not to disturb any damage to the firing sears I touched down smoothly without mishap. In true tradition, the ground crew creased themselves laughing as they wriggled my ejection pins in without disturbing the firing sear – the one delicately removing the detonation cord from around my ears was particularly amused. 'Your landing looked as gentle as an angel's fart sir, most unusual!'

Public Clearance day approached and Air Officer Training, AVM Clarke, cheerful as ever joined us. Having been the Chief Instructor at Lynton when I gained my wings he felt it only right to fly in a display practice with me to see if I had improved at all – and full credit to John it went well. The AVM's only comment was a team pilot putting his feet up, peeling an orange and offering some to him during a serious debrief was something new. John passed his comment to me with a huge grin.

1982 like the other seasons was fast, furious flat out and memorable. My last winter training at Kemble using Brize Norten left traces for a chuckle on another day. After barrel rolling around a new VC10 captain on a training flight (miles of clearance) and nipping in front to fly an inverted PAR approach, I was fascinated by his observation that breaking off at 100 feet upside down was a safety hazard. My advice to him was to give himself plenty of height rolling the right way up before beating up the tower. Got a chuckle for that. But no room for complacency, in summer we had a jolting reminder.

RAF Valley lost their first Hawk. Fumes in the cockpit of XX305 produced an acrid smell, the crew both went to 100% oxygen did the emergency drills and the instructor made an immediate return. The student had breathing difficulties, felt unwell and removed his mask. The instructor distracted over his condition unfortunately stalled on finals at 300 feet and ejected both of them but the student tragically didn't survive. The fumes were traced to the cold air unit and prompted a check throughout the fleet.

It was something to be taken seriously because the time to incapacitate was short. Byron our QFI went through drills and we pressed on. It was also the last season for Tim Watts who had raised the level of synchro to an art form and for Byron who ably steered us through with the Hawk: two great teammates with who I had flown umpteen shows. New guys replacing us joined us for a look-see in time for the Farnborough Air Show. Ted Ball fresh from the Falklands War, Curly Hirst and Simon Bedford (Goober) an old Hunter chum were seriously good pilots with personalities to match. But however good you are it's still a daunting experience to see a completely relaxed team suddenly cluster around the Boss. The brief is in a single sentence, startup up is precise and launch is exactly on the button and after hurling around with a dash of elan the team are back on the ground 35 minutes later to carry on a conversation as if nothing had happened in between. This is because Farnborough is in September towards the end of the season when the team is probably turning in their best performance for the public and at this point, you don't want to foul anything up and end the season leaving the public with a poor impression. Ejecting during a display would do that. And if you did that here the boss would bloody well kill you! And I almost considered doing it.

I was to fly with Simon in the back and it coincided with new masks being fitted to our helmets, these were washed in a strong ghastly smelling antiseptic that permeated your socks and soul. Trouble was the smell masked everything – Including the smell of a cold air unit failing because the grease around the bearings was the wrong sort. When low-speed grease is packed around a high-speed bearing it cracks and produces noxious fumes and Simon was having a ball in the back when the problem arose. We were nearing the end of the first half with several formation changes left before a break for synchro when things went hairy.

We both agreed the smell of the mask was like a gorilla's armpit when without warning my hands felt huge and cumbersome, far too big to fit around the stick which was quite alarming. Something was going pear-shaped for sure when my tongue began to feel as huge as my hands and my voice sounded thick. I twigged I was experiencing a variation of what happened to the crew in the Valley crash and told Simon to go to 100% oxygen pronto. He immediately did but I had a bit of difficulty taking my hand off the throttle being in tight formation so I continued to get an extra whiff of something. I knew that because my colour vision started to go and everything began to turn blue, which is bloody disconcerting when all the aeroplanes around you should be red!

Apart from keeping formation while the fun and games were going on, I had to get the air conditioning off and ram air through. Simon was steady as a rock – but out of it. It was a bit difficult with no chance of taking my hands off to do it, so I started yelling into my mask to keep the blood pressure up. It seemed to work, seconds later we were into the break, I flashed through my space, and hands-free promptly dumped the air-con, got ram air and prepared to bang out if everything went from blue to black. We were clear of the crowd so no sweat. Amazingly within 30 seconds, the air felt mountain pure, my tongue normal and hands the right size, and I must say I have never enjoyed a second-half of a display more.

The footnote on landing was the engineering guys on to it fast, I think the only cold air unit affected was ours and a signal flashed the incident around to shed more light on the problem. Bit of a bunchy experience but it worked out OK and I have to say I had a ball on my last Farnborough shows.

131 displays in my last season with John Blackwell leading brought home to me how much being part of the cutting edge matters in the RAF. Having so much fun since I joined as a boy getting there was remarkable luck but I had reached the point where I knew my future lay in active aviation: whether it be in the RAF perhaps with a command or test-flying or civilian big jets remained to be seen. To help my decision along to leave at age 38 or stay till 55, a senior desk officer from RAF Innsworth (sorts out your future appointment) dropped by one Friday to let me know what the future held. The only stumbling block to a future was my habit of brushing aside promotion paths that clashed with enjoyment, so I wasn't holding my breath.

But he blew my socks off!

A brief apology from the RAF for having to hold as a youngster on Shackletons waiting to train as a pilot broke the ice and raised a laugh. I assured him I was a beneficiary there. He read out the Navy assessment of my tour with them which was along the lines of endorsing me as a reasonable cove for joining the Royal Navy if I ever expressed a wish to do so. This came as a complete surprise considering the amount of time I spent taking the mickey out of them but I felt enormously flattered as I valued my time sharing their ethos.

But the big appointment was one I never knew existed. An exchange tour flying F14 Tomcats with the US Navy was firmly on the table, training to start at Miramar California and their fighter weapons school. My return tour would be as a flight commander on Air Defence Tornados and I would need to extend by three years to amortise. It bowled me over. Almost missed the promotion that went with it!

My natural urge was to fly a Tomcat and try to pull the wings off it beating US Navy pilots and staying with the RAF. But on the other hand, I had a burning desire to explore every facet of aviation. I heard Cathay Pacific Airways were interested in guys like me despite my lack of commercial experience, and the thought of hauling a Jumbo around the skies and facing the challenges of Kai Tak in lousy weather on three times the salary was not to be ignored.

I felt enormous gratitude to the RAF for such a great chance but bizarrely the offer was almost too good. What the hell was I to do? I needed time to think.

Was it going to be a punchy continuation of a job I was cut out for, or would I be risking everything just to explore new ways of walking the skies? Would I forever look skywards with envy when the time came to fly a desk or would I face new challenges flying the Pacific Rim in a big jet?

Only one way to find out! I went fly fishing with a lot on my mind and as always the answers floated gently to the surface …

Acknowledgements

My sincere thanks go to the RAF and Royal Navy for letting me fly Her Majesty's aeroplanes – most of which I returned in one piece. To Captain Pete John for encouraging me to put good times on paper and to Bruce Lawson, a writer and neighbour who convinced me someone out there might want to read about my howlers in flying as well as the heroes in aviation. I managed the right phraseology in Navy sequences thanks to Captain Peter Bowyer a near neighbour and a Bombardier and Boeing technical pilot who is ex-Royal Navy.

I have tried to credit the authors of all the photographs and images I have used. You have my sincere thanks. However, given the complexity of internet sourcing, it's no easy task to trace the true origin of some old photographs with any degree of certainty and if I have missed anyone please accept my apologies.

A special thanks to Michael Rondot for his advice and permission to use images of his famous paintings and to Claire Hartley for allowing me to use some remarkable contemporary photographs of the Reds. My thanks to imagery courtesy of James Biggadike.

Farewell to the Reds

ND - #0376 - 030123 - C51 - 229/152/18 - PB - 9781914424588 - Gloss Lamination